BURLEIGH DODDS SCIENCE: INSTANT INSIGHTS

NUMBER 101

Biofertiliser use in agriculture

I0053256

burleigh dodds
SCIENCE PUBLISHING

Published by Burleigh Dodds Science Publishing Limited
82 High Street, Sawston, Cambridge CB22 3HJ, UK
www.bdspublishing.com

Burleigh Dodds Science Publishing, 1518 Walnut Street, Suite 900, Philadelphia, PA 19102-3406, USA

First published 2024 by Burleigh Dodds Science Publishing Limited
© Burleigh Dodds Science Publishing, 2024, except the following: Chapter 3 was prepared by a
U.S. Department of Agriculture employee as part of their official duties and is therefore in the public
domain. All rights reserved.

British Library Cataloguing in Publication Data
A catalogue record for this book is available from the British Library

ISBN 978-1-80146-675-2 (Print)
ISBN 978-1-80146-676-9 (ePub)

DOI: 10.19103/9781801466769

Typeset by Deanta Global Publishing Services, Dublin, Ireland

Contents

Series list

Title	Series number
Sweetpotato	01
Fusarium in cereals	02
Vertical farming in horticulture	03
Nutraceuticals in fruit and vegetables	04
Climate change, insect pests and invasive species	05
Metabolic disorders in dairy cattle	06
Mastitis in dairy cattle	07
Heat stress in dairy cattle	08
African swine fever	09
Pesticide residues in agriculture	10
Fruit losses and waste	11
Improving crop nutrient use efficiency	12
Antibiotics in poultry production	13
Bone health in poultry	14
Feather-pecking in poultry	15
Environmental impact of livestock production	16
Sensor technologies in livestock monitoring	17
Improving piglet welfare	18
Crop biofortification	19
Crop rotations	20
Cover crops	21
Plant growth-promoting rhizobacteria	22
Arbuscular mycorrhizal fungi	23
Nematode pests in agriculture	24
Drought-resistant crops	25
Advances in detecting and forecasting crop pests and diseases	26
Mycotoxin detection and control	27
Mite pests in agriculture	28
Supporting cereal production in sub-Saharan Africa	29
Lameness in dairy cattle	30
Infertility and other reproductive disorders in dairy cattle	31
Alternatives to antibiotics in pig production	32
Integrated crop–livestock systems	33
Genetic modification of crops	34

Chapter 1

Organic fertilizers and biofertilizers

Lidia Sas Paszt and Slawomir Gluszek, Research Institute of Horticulture, Poland

1 Introduction

Organic fertilizers contain organic substances, including amino acids, sugars, lipids, vitamins, enzymes, plant hormones, macro- and microelements and other active compounds, which stimulate plant growth and development (Crouch and van Staden, 1993; Stirk and Staden, 1996; Khan et al., 2009; Ji et al., 2017).

Products permitted for use in organic farming in the European Union (EU) are listed in 'Annex I' to the EEC Regulation 2092/1991. At present, there is 'Annex II' – 'Products for use in fertilization and soil conditioning' (Løes et al., 2016). This list covers a wide range of products, from manures and composts to sediments from freshwater bodies. All of them are used as raw materials or, nowadays, as highly processed, modern products in the form of concentrated liquids, enriched powders, granulates or similar products. New bio-based formulations make it possible to save energy, labour input and time needed for efficient field application and plant nutrition (Vaneeckhaute et al., 2013). Biofertilizers contain, among other components, living cells of microorganisms

http://dx.doi.org/10.19103/AS.2021.0094.18

as the biologically active ingredient. The products suitable for organic agriculture that are available on the European market include a large group of products ranging from dried manures to highly processed organic-mineral fertilizers and mycorrhizal and bacterial inocula. Despite numerous studies on the use of preparations of organic origin, there is still a lack of bio-stimulants and protective substances for enriching the habitats of plants microbiologically. Further work is necessary on improving the technologies of cultivation and fertilization of fruit plants.

2 Biofertilizers

There has been, so far, no precise definition for the term 'biofertilizer'. Some authors define it as a substance that contains, as an active ingredient, live cells of microorganisms separately or in combination with other active ingredients that colonize the rhizosphere or spaces inside plant tissues and that help in enhancing the nutrient uptake by plants, and is thus able to enrich the nutrient quality of the soil (Vessey, 2003). This definition limits biofertilizers to microbes that can assimilate free nitrogen or dissolve water-insoluble salts of phosphorus, potassium or other nutrients. Sometimes, the term 'biofertilizer' is used for artificially multiplied microbial inoculants that can improve soil fertility and crop productivity (Roychowdhury et al., 2015).

Others define biofertilizers as biologically active products containing organic compounds such as amino acids, sugars, vitamins and other substances that have a direct nutritional impact on plant growth and yielding.

However, limiting the action only to nutritional effects means that in many cases, there is a problem with a wide group of organic products (both microbial and of organic origin) which also have other effects, for example, protection against diseases or pests or plant growth regulation.

A large group of compounds for agriculture, especially for organic farming, that contain living microbial cells are called biofertilizers. This category covers products that contain live cells of bacteria, cyanobacteria, actinomycetes, yeast cells and spores and/or hyphae of microscopic, filamentous or mycorrhizal fungi. The bacterial and other microbial strains used in the products for organic agriculture belong to various genera and species.

The mechanisms by which plant-growth-promoting rhizobacteria (PGPR) enhance plant growth are not fully understood. However, it is believed that the PGPR promote plant growth and yielding by either direct or indirect mechanisms, where direct mechanisms include the ability to produce phytohormones like indole acetic acid (IAA), gibberellin, cytokinins and ethylene; asymbiotic N_2 fixation; antagonism against phytopathogenic microorganisms by the production of siderophores; and solubilization of mineral phosphates and other nutrients; and indirect mechanisms entail the extracellular production of antibiotics, synthesis of antifungal metabolites, production of fungal cell wall lysin enzymes, depletion of iron from the rhizosphere and competition for sites on roots and induced systemic resistance (Salantur et al., 2006; Egamberdiyeva, 2007; Ahmad et al., 2008; Aznar et al., 2014; Gusain et al., 2015).

2.1 Nitrogen-fixing microorganisms

'Traditional' biofertilizers contain symbiotic bacteria, for example, rhizobia (*Rhizobium, Neorhizobium, Allorhizobium, Pararhizobium, Mesorhizobium, Bradyrhizobium* and *Ensifer*), free-living bacteria, such as *Azotobacter, Azospirillum*, or cyanobacteria like *Nostoc, Anabaena, Tolypothrix* or *Aulosira*, which are able to fix free nitrogen from the air (Souza et al., 2014; Dhar et al., 2015).

2.1.1 Rhizobia

Rhizobia are bacteria that formerly belonged to the family Rhizobiaceae, but at present, some of them are classified into new families: Phyllobacteriaceae into the genus *Mesorhizobium*, Nitrobacteriaceae (Bradyrhizobiaceae) into the genus *Bradyrhizobium* and Burkholderiaceae into the genus *Paraburkholderia* (formerly *Burkholderia*) (Velázquez et al., 2017). The complete list of valid species of rhizobia is constantly updated and recorded as the 'List of Prokaryotic names with Standing in Nomenclature' (http://www.bacterio.cict.fr).

These bacteria are mutualistic symbionts, and fix nitrogen environmentally depended highly variable at about 50–350 kg/ha, with legumes only. They are useful for pulse legumes like chickpea, red-gram, pea, lentil, black gram and so forth; oilseed legumes like soybean and groundnut; and forage legumes like berseem, clovers and lucerne. Successful nodulation of leguminous crops by a *Rhizobium* largely depends on the availability of compatible strains for a particular legume. Rhizobia colonize the roots of specific legumes to form tumour-like growths called root nodules, which act as factories of ammonia production. Rhizobia have the ability to fix atmospheric nitrogen in a symbiotic association with legumes and certain non-legumes, like *Parasponia*. Each legume requires a specific species of *Rhizobium* to form effective nodules. Many legumes may be nodulated by diverse strains of Rhizobia, but growth is enhanced only when nodules are produced by the effective strains of Rhizobia. It is thus extremely important to match microsymbionts prudently for maximum nitrogen fixation, especially because some native strains are not as effective as host-specific rhizobial strains, and there is a possibility of establishing symbiosis with wild 'cheating', non-effective nitrogen-fixing strains (Kiers et al., 2003; Vicario et al., 2016). A strain of *Rhizobium* that nodulates and fixes a large amount of nitrogen in association with one legume species may also do the same in association with certain other legume species. Leguminous plants demonstrate the tendency to respond similarly to particular strains of rhizobia (Wani and Lee, 2002).

The population of *Rhizobium* in the soil depends on the presence of legume crops in the field. In the absence of legumes, its population decreases. Artificial seed inoculation is often needed to restore the population of effective *Rhizobium* strains near the rhizosphere to hasten N_2-fixation. Symbiotic associations of legumes and rhizobia can be a major source of nitrogen in organic and sustainable systems, whereas free-living nitrogen-fixing microorganisms are helpful in non-symbiotic systems (Rashid et al., 2016a).

2.1.2 Azospirillum

Azospirillum belongs to the family Spirillaceae and is heterotrophic and associative in nature. It has nitrogen-fixing ability of about 20-40 kg/ha. Besides nitrogen fixation, bacteria from the genus *Azospirillum* are also able to produce phytohormones: auxins, gibberellins, cytokinins and phytohormone-like agents, such as indole derivatives (Cassán et al., 2014). Although there are many species under this genus, such as *Azospirillum amazonense*, *Azospirillum halopraeferens* and *Azospirillum brasilense*, the worldwide distribution and benefits of inoculation have been proved mainly with the *Azospirillum lipoferum* and *Azospirillum brasilense*. *Azospirillum* forms associative symbiosis with many plants, particularly with those having the C4-dicarboxylic pathway of photosynthesis (the Hatch and Slack pathway), because they grow and fix nitrogen on salts of organic acids such as malic and aspartic (Arun, 2007). Thus, they are mainly recommended for maize, sugarcane, sorghum, pearl millet and so forth, especially in environmental stress conditions like drought (Moutia et al., 2010). *Azospirillum* colonizing the roots do not only remain on the root surface, but a sizable proportion of them also penetrate into the root tissues and live in harmony with the plants. They do not, however, produce any visible nodules or outgrowth on root tissue.

2.1.3 Azotobacter

Azotobacter belongs to the family Azotobacteraceae, and is aerobic, free living and heterotrophic in nature. *Azotobacter* are present in neutral or alkaline soils, with *Azotobacter chroococcum* being the most commonly occurring species in arable soils. *Azotobacter vinelandii*, *Azotobacter beijerinckii*, *Azotobacter insignis* and *Azotobacter macrocytogenes* are other reported species. The number of *Azotobacter* rarely exceeds 10^4-10^5 g^{-1} of soil due to lack of organic matter and presence of antagonistic microorganisms in the soil. The bacterium produces antifungal antibiotics which inhibit the growth of several pathogenic fungi in the root region, thereby preventing seedling mortality to a certain extent (Nagaraja et al., 2016). An isolated culture of *Azotobacter* fixes about 10 mg nitrogen g^{-1} of carbon source under *in vitro* conditions. *Azotobacter* are also known to synthesize biologically active growth-promoting substances such as vitamins of the B group, IAA and gibberellins. Many strains of *Azotobacter* also exhibit antifungal properties against plant pathogens such as *Fusarium* or *Alternaria* (Chetverikov and Loginov, 2009). They can also be applied in environmental stress management (Di Stasio et al., 2017). The population of *Azotobacter* is generally low in the rhizosphere of crop plants and in uncultivated soils. The occurrence of this organism has been reported in the rhizosphere of a number of crop plants such as rice, maize, sugarcane, vegetables and plantation crops (Arun, 2007).

2.1.4 Blue-green algae (Cyanobacteria)

Blue-green algae belong to eight different families and are phototrophic in nature, producing auxin, IAA and gibberellic acid, fixing about 20–30 kg N/ha in submerged rice fields. Since they are abundantly found in paddies, they are also referred to as 'paddy organisms'. N is the key input required in large quantities for lowland rice production. Soil N and biological nitrogen fixation (BNF) by associated organisms are major sources of N for lowland rice. The 50-60% N requirement is met through the combination of mineralization of soil organic N and BNF by free-living and rice-plant-associated bacteria (Roger and Ladha, 1992). To achieve food security through sustainable agriculture, the requirement for fixed nitrogen must be increasingly met by BNF rather than by industrial nitrogen fixation. Most N_2-fixing BGA are filamentous, consisting of a chain of vegetative cells, including specialized cells called heterocysts which function as micronodules for the synthesis and N_2-fixing process. BGA forms a symbiotic association capable of fixing nitrogen with fungi, liverworts, ferns and flowering plants. The most effective common symbiotic association has been found for the *Azolla* and *Anabaena azollae* (BGA) complex. Besides N_2 fixation, BGA are also capable of producing siderophores which can complex metal ions into stable forms (Singh et al., 2016).

   Strains of other free-living and endophytic bacterial species are also capable of fixing nitrogen from the air (Fox et al., 2016). Nitrogen-fixing microorganisms seem to be an economically attractive and ecologically friendly alternative to artificial nitrogen fertilizers, especially in sustainable agricultural production.

2.2 Non-nitrogen-fixing bacteria

Non-nitrogen-fixing bacteria are also suitable for agricultural use both as biofertilizers and plant growth promoters, especially as nutrient (e.g. phosphate) solubilizers, which are able to solubilize inorganic phosphate compounds, such as tricalcium phosphate, dicalcium phosphate, hydroxyapatite and rock phosphate. Among the bacterial genera with this capacity are *Pseudomonas*, *Bacillus*, *Rhizobium*, *Burkholderia*, *Achromobacter*, *Agrobacterium*, *Micrococcus*, *Aerobacter*, *Flavobacterium* and *Erwinia*. There are considerable populations of phosphate-solubilizing bacteria in the soil and the plant rhizosphere. These include both aerobic and anaerobic strains, with a prevalence of aerobic strains in submerged soils. A considerably higher concentration of phosphate-solubilizing bacteria is commonly found in the rhizosphere in comparison with the bulk soil (Roychowdhury et al., 2015). The soil bacteria which belong to the genera *Pseudomonas* and *Bacillus*, and fungi are more common. The major microbiological means by which insoluble P compounds are mobilized is by the production of organic acids, accompanied by acidification of the micro-environment (Raghu and MacRae, 1966). The organic and inorganic acids convert tricalcium phosphate to di- and monobasic phosphates with the net result of an enhanced availability of the element to the plant. The type of organic acids produced and their amounts differ with different organisms (Chen et al., 2006). Tri- and di-carboxylic acids are more effective compared to monobasic and aromatic acids. Aliphatic acids are also found to be more effective in P solubilization compared to phenolic, citric and fumaric acids (Walpola and Yoon, 2012). The analysis of culture filtrates of phosphate-solubilizing microorganisms has shown the presence of a number of organic acids, including citric, fumaric, lactic, 2-ketogluconic, gluconic, glyoxylic and ketobutyric acids (Wani et al., 2015).

Other rhizosphere microorganisms can reduce manganese and iron into water-soluble forms which are available for uptake by plants. In particular, bacteria belonging to the genera *Bacillus*, *Pseudomonas*, *Citrobacter*, *Shewanella*, *Alcaligenes*, *Clostridium* and *Enterobacter* are capable of reducing insoluble salts of iron and manganese (Ebrahiminezhad et al., 2017). The reduction of metal oxides in the soil can be linked to anaerobic oxidation of methane, one of the important greenhouse gases in low-oxygen soil conditions (Oni and Friedrich, 2017). Some beneficial bacteria, like *Pseudomonas*, *Rhizobium*, *Azotobacter* or *Erwinia* sp., can produce chelating agents (siderophores) that can form soluble complexes with soluble ions, preventing them from reprecipitation (Born et al., 2016; Kumar et al., 2017). The production of chelating agents is also one of the mechanisms of pathogen suppression by beneficial microorganisms (Sasirekha and Srividya, 2016).

2.3 Mycorrhiza

Another important group of beneficial microorganisms are mycorrhizal fungi. In agriculture and horticulture, the most common are arbuscular mycorrhizal fungi (AMF) belonging to Glomeromycota. They are obligatory symbionts of plants and form symbiosis with almost 80% of plant species on the earth (Davison et al., 2015; Öpik et al., 2010). These fungi are associated with the majority of agricultural crops, except the crops/plants belonging to the families of Chenopodiaceae, Amaranthaceae, Caryophyllaceae, Polygonaceae, Brassicaceae, Commelinaceae, Juncaceae and Cyperaceae. AMF can produce a network of extraradical mycelia, which gives access to a greater quantity of water and soil minerals not available to the roots of the host plants. AMF also produce the peptide-like substance – glomalin, which plays a key role in the stabilization of soil structure and quality (Singh et al., 2013; Bedini et al., 2010; Vasconcellos et al., 2016). The fungi receive the necessary carbohydrates from plants for the completion of their life cycle. Arbuscular mycorrhizal associations have been shown to reduce damage caused by soil-borne plant pathogens. It is well documented that the arbuscular mycorrhizal symbiosis can increase plant growth and nutrient uptake, improve fruit quality and alleviate several abiotic stresses such as low temperature stress, drought, salt stress and so forth (Cavagnaro et al., 2015; Rouphael et al., 2015; Taktek et al., 2015; Zhao et al., 2015; Yooyongwech et al., 2016).

Application of microorganisms can increase the bioavailability of nutrients and the stability of soils and can support formation of soil aggregates in poor soils (Rashid et al., 2016b).

3 Consortia of microorganisms to improve the effectiveness of organic fertilization

Application of selected microorganisms may increase the effectiveness of organic fertilizers in plant production. Especially effective is co-application of microorganisms with organic fertilizers.

Microorganisms can be applied alone, as a single strain, but in many cases, better effects are observed after application of microbial consortia. The increased effects are achieved through the synergic relation between different groups of microorganisms. Application of consortia increases the nutritional effect of organic fertilizers, with additional 'value added' contributions, such as the production of phytohormones, antimicrobial agents, siderophores and others. Pseudomonads applied with AMF stimulate flower production and increase the yielding of strawberry, increasing yield parameters and amounts of biologically active substances in strawberry fruits (Bona et al., 2015). Microbial consortia are helpful in better utilization of organic or mineral fertilizers.

Grzyb et al. (2015) have shown that the application of the commercially available microbial consortium 'Mycosat' with different organic preparations and fertilizers increased plant growth in sour cherry 'Debreceni Bötermö' and apple 'Topaz' than the application of fertilizers alone. 'Mycosat' stimulated the growth of plants and promoted better bud-take on inoculated rootstocks. The resistance of bud grafts to freezing was increased. A disadvantage, however, is its relatively low effectiveness when the soil has become excessively dry. A double application of this preparation during the growing season gave better results in the nursery than a one-off treatment.

Common applications of AMF (*Rhizophagus intraradices*, *Glomus aggregatum*, *Glomus viscosum*, *Claroideoglomus etunicatum* and *Claroideoglomus claroideum*) and plant-growth-promoting *Pseudomonas* strains (C7 and 19Fv1T) positively affected the flowering and fruit production, and also the amounts of vitamin C, sugars and organic acids in tomato plants cv. 'TC 2000' (Bona et al., 2017).

Application of a mixture of beneficial bacteria namely *Pantoea* sp., *Pseudomonas fluorescens*, *Klebsiella oxytoca* and *Rhizobium* sp. on organically or mineral-NPK fertilized fields caused better plant growth and yielding of apple cv. 'Topaz' in comparison with application of single mineral or organic fertilizers alone (Mosa et al., 2016). Preparations with AMF phosphate-solubilizing bacteria and *Azotobacter* combined with vermicompost increased the growth and yielding of strawberry plants in comparison with vermicompost single application (Singh et al., 2015). Similar effects were observed in basil (*Ocimum basilicum*) after co-application of vermicompost, *Dietzia natronolimnaea* and *Glomus intraradices* (Bharti et al., 2016b).

The synergic effect of different microorganisms is taken advantage of in the production of commercial products which contain beneficial bacterial strains and arbuscular or filamentous fungi, sometimes mixed with rock meal as a source of phosphorus, potassium and other minerals (Grzyb et al., 2015).

A group of products called 'Effective Microorganisms' (EM), popular in organic agriculture, is obtained during molasses fermentation (see Section 6.1) using a consortium of many microbial strains. EM, when applied as fertilizer, resulted in positive impacts on plant growth parameters (Kleiber et al., 2014; Salama et al., 2014). For example, strawberry plants cv. 'Honeoye' treated with EM gave similar results to those under mineral fertilization, whereas cv. 'Selva' showed better growth parameters after the application of 'Effective Microorganisms' (Glinicki et al., 2011). Indirect, positive effects in plant growth and yielding are obtained when 'Effective Microorganisms' are used as microbial inoculum for compost preparation. For example, in an eleven-year long-term experiment, wheat fertilized with 'Effective Microorganisms' compost gave better plant growth parameters and yield in comparison to traditional compost (Hu and Qi, 2013). 'Effective Microorganisms' are especially useful in co-application with organic matter like agricultural wastes or with bioorganic fertilizers while increasing the effectiveness of nutrients utilization (Chantal et al., 2013; Shaheen et al., 2017).

4 Animal excrement: manures, slurry and guano

The most common organic fertilizer used in organic farming is farmyard animal manure, which is the basis of organic fertilization of many crops. In recent years, there has been a problem with adequate amounts of this resource for both organic and integrated/conventional agriculture. Manure can be applied in fresh, dried or composted form. However, for safe organic production, to avoid the risk of contamination with pathogenic or opportunistic strains of microorganisms, it is recommended that manure be thermally processed, for example, during thermophilic composting. If raw manure is used, it should be applied 90 days before harvest for crops that do not come into direct contact with the soil, and 120 days prior to harvest for crops that have direct contact with the soil (Köpke et al., 2007). The use of manure originating from factory farming is forbidden.

Long-term application of farmyard manure increases the levels of P, K and S available to plants, and also the amounts of microelements like B, Zn, Fe or Mo in the soil, but can decrease the concentration of Cu and Mn (Rutkowska et al., 2014). Application of manures modifies the amounts of available forms of nutrients in the soil profile. This effect is dependent on the origin and form of the applied manure (Schlegel et al., 2017).

Manures are a source of soluble substances which are easily utilized by soil microorganisms. These substances are sugars, alcohols and volatile fatty acids such as acetic, butyric and valeric acids. Slowly biodegradable organic polymers must be hydrolysed before their utilization by soil microorganisms (Boursier et al., 2005). At present, manures are available on the market in processed form, dried or composted or as dried granules or pellets. Some of these products are supplemented with rock powders, animal blood, bone or feather meals, plant by-products or other materials (Malusà et al., 2014).

Other traditional products of animal origin, widely used as fertilizers, are slurry (liquid animal excrement) used after controlled fermentation and/or appropriate dilution.

Guano, accumulated excrement of seabirds, seals or cave-dwelling bats, is a highly effective fertilizer with high nutrient contents essential for plant growth. However, its mineral content is strictly associated with the feeding regimes of guano-producing animals (Szpak et al., 2012). For example, the guano produced by the sanguivorous bat *Desmodus rotundus* has more than 17% nitrogen, that produced by the frugivorous bat *Pteropus rodricensis* has about 2% and the guano produced by the insectivorous bat *Tadarida brasiliensis* has a nitrogen content of about 10% (Emerson and Roark, 2007). Large deposits of guano are formed in dry climates, where they are protected from nitrogen leaching by rain.

5 Products and by-products of animal origin

5.1 *Raw or processed animal residues*

A large group of organic fertilizers are raw or processed animal residues, dried and pulverized blood, hoof, horn, bone, degelatinized bone, fish, meat, feather, hair, 'chiquette' meal,'chiquette', wool, fur, hair, dairy products and hydrolysed proteins.

Animal residues are sources of organic matter and minerals N, P, Ca and others. But the usage of these materials, especially meat and bones, for agriculture is restricted by safety rules, particularly those relating to TSE (transmissible spongiform encephalopathy) incidents (Möller, 2015). Substrates allowed for the preparation of organic fertilizers without pressure sterilization are those classified to Category 3 according to the EC Regulation 1069/2009. By-products in this category are obtained from the processing of goods intended for human consumption obtained by slaughtering animals, leftovers from food processing facilities or canteens. This category of by-products, when used for the preparation of organic fertilizers, must be treated by pasteurization at 70°C for more than 1 h. Category 2 covers by-products from animals that died in ways other than slaughtering, animals killed to prevent epizootic diseases or those posing a risk of infection with diseases other than TSE. This category also covers the contents of digestive tracts obtained by slaughtering healthy animals. These residues may be used for the production of fertilizers after pressure sterilization at a temperature of more than 133°C for more than 20 min under a pressure above three bars. Processing of meals into fertilizers may be done by simple granulation to obtain a solid phase. Category 2 by-products can be processed by acid hydrolysis to obtain short peptide chains suitable for the production of solid nitrogen-rich fertilizers. For example, 'Biollsa', prepared by Ilsa Group, Italy, is a granular nitrogen fertilizer produced by acid hydrolysis.

Another group of fertilizers of animal origin, popular in recent years, are liquid hydrolysates rich in amino acids and short peptide chains. The best of these preparations are obtained by enzymatic hydrolysis of purified animal residues such as collagen. For maximum safety, some producers use only Category 3 residues to avoid the risk of spreading disease. Liquid formulation makes them easy to use by spraying, especially in pomological orchards, berries plantations, vineyards and fields. Liquid hydrolysates are also used via irrigation systems, in greenhouse production of high-quality vegetables and fruits. Hydrolysed proteins must not be applied to edible parts of the crop, and in the case of pastures, the feeding of animals must be prohibited for at least three weeks to avoid the potential risk of disease infection. Amino acids are also a source material for the production of complexes with micronutrients, which are eco-friendly plant growth promoters (Ghasemi et al., 2013). The effectiveness of amino acid preparations in plant nutrition and for improving soil quality has been investigated in many trials, which have shown that applications of such preparations increase the yield quantity and quality (Lisiecka et al., 2011; Corte et al., 2014; Colla et al., 2015; Mobini et al., 2014). For example, the Terra-Sorb Foliar amino acid preparation applied on oilseed rape increases the levels of fats and proteins and decreases the amounts of glucosinolates in seeds in comparison with the control (Jakienė, 2013).

Wool, fur and hair are a source of slowly released nitrogen when used in a low-processed form like pellets or meal. They are also used as a feedstock for the production of more processed preparations such as hydrolysates (Nustorova et al., 2006).

Dairy products and by-products, especially whey, have been a point of interest as a form of nitrogenous fertilizer for a long time (Sharratt et al., 1959; Peterson et al., 1979; Marwaha and Kennedy, 1988). Direct application of whey on arable fields could be a method for improving soil fertility and for improving the physical properties of some types of degraded soils (Robbins and Lehrsch, 1992; Jones et al., 1993).

Good results of whey utilization are obtained when it is used as a raw material for organic fertilizer production. Whey-based fertilizers are compositions of different organic amendments like sawdust and straw, and also with minerals like zeolite. An example of such a product is 'Condit' – a pelletized sawdust, straw and zeolite mixture enriched with whey as a source of nitrogen (Cuijpers and Hospers-Brands, 2008). Whey is also a good supplement for bioconversion of some substrates of organic origin into valuable organic fertilizers that have positive effects on plant growth parameters (Stępień et al., 2012). They are also an effective amendment for compost teas, especially in the protection against plant pathogens (Pane et al., 2012).

5.2 Chitin

Chitin fertilizers are obtained from shells of crustaceans. This product is allowed only if obtained from sustainable fisheries, as defined in EC 2371/2002, or organic aquaculture. Chitin and its derivatives are not only sources of nitrogen and carbon, but also active against viruses, bacteria and other pests; they also induce plant defence mechanisms and stimulate the growth and activity of beneficial microorganisms (Schäfer et al., 2012; Cretoiu et al., 2013; Sharp, 2013), and can also increase the effectiveness of beneficial bacteria against pests (Yu et al., 2008). Chitin fragments are known to have eliciting activities, leading to a variety of defence responses in host plants. Based on these and other properties that help to strengthen host plant defences, interest has been growing in using the technique in agricultural systems to reduce the negative impact of diseases on the yield and quality of crops. When applied as a foliar spray, 0.1% (w/v) chitosan solution increases turmeric plant (*Curcuma longa* L.) growth parameters, such as shoot height, number of leaves per plant and plant fresh weight, and also has a positive impact on plant yielding and yield quality, including active substances content (Anusuya and Sathiyabama, 2016). Chitosan is especially effective when used in combination with microelements and humic acids, increasing, for example, dry bean (*Phaseolus vulgaris* L.) plant growth parameters and yield quality (Ibrahim and Ramadan, 2015). Recently, positive results with chitin have been obtained in a study conducted by Mahdavi and Rahimi (2013). They tested the effect of 'Chitosan' on the germination and growth of ajowan (*Trachyspermum ammi*). Stimulation of germination and growth was observed with a decrease in the harmful impact of abiotic stress such as high salinity. Significant benefits have been observed for soybean yield, seed germination and plant growth in a study by Zeng et al. (2012), in which soybeans were coated with chitosan. These properties make chitin and its derivatives a useful material for coating seeds used in organic agriculture.

Chitin is also a raw material for the production of innovative liquid organic fertilizers, which are easy to use and effective when applied to plants and soil (Struszczyk et al., 1989). Innovative formulations, like the preparation 'Apol-Humus' allowed for use in organic agriculture in Poland, are a liquid mixture of chitosan and humic acids for foliar and soil application. This preparation stimulates the growth of plants in two ways: through a direct nutritional effect and disease suppression.

6 Products and by-products of plant origin for fertilizers

The 'products and by-products of plant origin' category covers a wide group of leftovers from the food processing industry, for example, oilseed meals, cocoa husks and malt culms. These by-products are sources of slow releasing nitrogen suitable for use in agriculture and horticulture. The levels of nutrients, especially of nitrogen forms, vary according to the source of the raw material and the extent of processing. For example, the average nitrogen content in sunflower meals is about 5.0%, *Brassica* meal 5.7%, neem (*Azadirachta indica*) cake 5.2%, castor cake 4.25% and cotton seed cake 4.6% (Kumar et al., 2014; Mazzoncini et al., 2015). In the case of meals obtained from seeds of some *Brassica* species, there are high levels of glucosinolates, which have an inhibitory effect on nitrification processes (Mazzoncini et al., 2015). The nutritional effect of plant seed meals can be enhanced due to their additional properties, for example, antifungal or zoocidal activity, especially against root-knot nematodes such as *Meloidogyne* (Duong et al., 2014; Sumbul et al., 2015; Yang et al., 2015a).

For fertilization purposes, processed residues of many plant species, especially legumes, are also used. For example, pelletized meal of red clover (Ekofert K) gave the highest marketable yield of celeriac cv. 'Diamant' in comparison with traditional mineral fertilization and non-fertilized control (Kaniszewski et al., 2013).

6.1 Molasses

Another plant product, molasses, is a raw material for the production of some soil improvers and fertilizers. For example, Vinassa is obtained as a by-product during the production of baker's yeast. Vinassa, when applied as a foliar spray, positively affects plant growth and yielding, but its effects are associated with the plant species, variety and also growing conditions such as the type of soil or field localization. For example, large-fruited cranberry (*Vaccinium macrocarpon* L.) cv. 'Stevens' shows good results in terms of plant growth parameters in comparison to 'Pilgrim' and 'Ben Lear' after applications of Vinassa (Derkowska et al., 2015b). In nursery stock production, Vinassa, applied at a rate of 5 L/ha, moderately stimulates the growth of maiden trees, increasing their height and the length of lateral shoots, and has a beneficial effect on the development of the root system and flower buds. Apple trees cv. 'Ariva' produced the best quality fruits as a result of treatment with Vinassa (Rozpara et al., 2014). Also soil microbial community parameters, such as the formation of mycorrhizal structures and the population size of free-living bacteria and fungi communities, positively affected after treatments with Vinassa (Derkowska et al., 2015a; Sas Paszt et al., 2015).

6.2 Natural waste fibres

Natural waste fibres are used as raw materials of plant origin from the textile industry. New products based on these materials are also obtained. These materials are used as a slow release fertilizer when mixed with a source of nitrogen, for example, residues of legume plants. For example, they can be used for the production of agro-fleece for organic vegetable production, which protects crops against weeds, and after crop harvest, the fleece is tilled under and acts as a slow release fertilizer and soil structure improver (Babik et al., 2013). In that case, agro-fleece was enriched with clover leaves as a source of nitrogen, and the obtained yield was similar to that obtained using mineral plant nutrition, but with minimal weed growth in the covered fields.

6.3 Plant-derived hydrolysates

Plants are a source of raw material for the production of plant-derived hydrolysates used in agriculture as bio-stimulants. Hydrolysates of plant origin show not only a nutritional effect but also a hormone-like activity, which results in higher biomass, SPAD index and nutrient content in the tissues of the treated plants (Colla et al., 2014).

'BioFeed Amin', a product based on plant hydrolysates, applied in a dose of 5 L/ha, intensifies the growth of maiden trees, markedly improves their branching ability, stimulates the growth of lateral shoots and improves considerably the quality of the root system. Application of plant-derived hydrolysates has a positive impact on soil microbial communities in comparison with mineral fertilization regimes, resulting in higher number of bacteria in the soil and better mycorrhiza formation in the roots (Sas Paszt et al., 2011; Derkowska, et al., 2015a).

6.4 Seaweeds and seaweed products

Seaweed and seaweed-based products have been used in agriculture for a long time, first as plant material directly applied to the soil, later as solid meals used as fertilizers or soil improvers, with their mode of action based on the minerals released into the soil (Crouch and van Staden, 1993). Seaweeds are a rich source of natural compounds such as amino acids, polysaccharides, lipids, trace elements, vitamins and especially plant growth regulators (auxins, cytokinins, gibberellins, betains, etc.), making seaweed suitable for sustainable, especially organic, agriculture. Seaweed products affect plant growth and biotic and abiotic stress resistance when applied directly onto plants and stimulate the rhizosphere microorganisms when applied to the soil (Crouch et al., 1990; Kuwada et al., 2006; Thorsen et al., 2010; Panda et al., 2012). Seaweed preparations directly affect plant growth and yield quality, both in agricultural and horticultural production (Fan et al., 2011; Hernández-Herrera et al., 2014; Kumar and Sahoo, 2011; Papenfus et al., 2013). In strawberry, for example, the effects of seaweed preparations Kelpak® and Goëmar BM 86® interacted with the plant variety. In the case of 'Elkat', the preparations Kelpak SL and Goëmar BM 86® significantly improved fruit yield, but in the case of 'Salut', there was no effect on yield (Masny et al., 2004). The seaweed preparations Kelpak® and Goëmar BM 86® applied to apple trees, on their own or together in a mixture, stimulated the growth of shoots and leaves and modified fruit coloration, but the effects were also dependent on the cultivar (Basak, 2008). In the case of agricultural crops, applications of the preparation 'Maxicrop' increased shoot and root dry weight of radish in comparison with another seaweed-based product – 'Seasol', and ashes obtained from these products, which were not as effective as Maxicrop (Yusuf et al., 2016). Seaweed extracts can be used to control pests on plants via direct effects or by increasing plant resistance (Hankins and Hockey, 1990; Jayaraj et al., 2008; Jayaraman et al., 2011; Stadnik and Freitas, 2014; Ngala et al., 2016). In the tests conducted in a nursery of the Research Institute of Horticulture in Skierniewice, Poland, the seaweed-based preparation 'BioFeed Quality', applied twice per season at 5 L/ha, intensified the growth of maiden trees, markedly stimulated their branching, caused a significant increase in the length of lateral shoots and had a beneficial effect on the development of the root system of the maiden trees.

In organic agriculture in the EU, only marine seaweed products obtained directly by physical processes including dehydration, freezing, grinding, extraction with water or aqueous acid and/or alkaline solutions, or fermentation can be used. The conditions of extraction have a significant impact on the physical and chemical properties of the final product and its effect on plants under cultivation conditions (Briceño-Domínguez et al., 2014).

6.5 Sawdust and wood chips

Sawdust and wood chips obtained from chemically untreated wood are a soil improver used especially in horticulture. Sawdust, wood chips, wood bark and other similar residues are commonly used as mulch in the cultivation of different horticultural plants, small fruits, vegetables and ornamentals (Haynes and Swift, 1986; Sanderson and Cutcliffe, 1991; Abouziena and Radwan, 2015; Lima et al., 2016). Application of these materials prevents the growth of weeds, but simultaneously 'starves' the cultivated trees of nitrogen, causing their leaves to become yellowish, so that in the following year, the trees grow less vigorously as the nitrogen content in the soil decreases compared to the control (Hoagland et al., 2008).

Another use of sawdust is as a medium or component of media for soilless production of plants, especially in protected (under cover) cultivation, being a promising alternative to the traditional growing media such as rockwool, peat or coconut fibre (Jung et al., 2015; Depardieu et al., 2016). Results of hydroponic production of strawberry fruit on sawdust show that the yield obtained on pure sawdust is lower than that obtained on media based on sawdust mixed with pumice or with coconut fibre (Marinou et al., 2013). Strawberry plants planted in sawdust-based media need additional fertilizer supplementation to satisfy the initial nutrient requirements of bare-root plants (Depardieu et al., 2016).

Sawdust and other wood derivatives are also suitable amendments for manure composting, for the production of composted fertilizers or soil improvers allowed for use in organic agriculture (Cuijpers and Hospers-Brands, 2008; Troy et al., 2012; Khan et al., 2014). Moreover, sawdust and wood chips are used for biochar production (Spokas et al., 2009; Ghani et al., 2013; Laghari et al., 2016).

6.6 Wood ash as fertilizer and soil improver

Wood ash used in organic agriculture must be obtained from wood that has not been treated with chemicals (preservatives, paints, etc.) after felling because this can negatively affect soil quality and plant growth (Jones and Quilliam, 2014). Wood ash contains all the essential plant nutrients, except nitrogen. Wood ash may be directly used to improve crop production and improve soil pH. Wood ash as a natural mineral fertilizer can induce negative priming and promote soil organic matter from grassland soil (Reed et al., 2017). Wood ash reduces fertilization costs and improves crop yields and soil properties, with residual effects observed for up to seven years or more. Wood ash is a good improver during composting of organic matter. Additions of up to 15% do not affect the composting process, but high doses of ash can increase the input of heavy metals present in the ash to limits that are not acceptable for good-quality composts (Fernández-Delgado Juárez et al., 2015).

6.7 Mushroom culture wastes (spent mushroom substrates – SMS)

The composition of this category is limited to products listed in 'Annex II' – 'Products for use in fertilization and soil conditioning'. A typical mushroom substrate is prepared by using straw, poultry litter or horse manure with the addition of gypsum, fermented and inoculated with fungal mycelia. This method is typical for *Agaricus bisporus* cultures. Other cultivable mushrooms, such as *Pleurotus* sp., are cultivated on wood or on cereal straw (Roy et al., 2015). These two kinds of spent substrates differ in nutrient content and physical properties (Paredes et al., 2016). Mushroom cultivation is one of the methods for the utilization of organic residues. However, large amounts of spent substrates are a serious problem for their utilization (Das and Mukherjee, 2007; Koutrotsios et al., 2014). They can be used as raw material in the field, or can be processed to obtain products with better effectiveness.

Spent mushroom substrates or their derivatives promote the growth and yields of plants and increase crop quality. For example, the growth of pepper plants was strongly induced by *Pleurotus* SMS leachate, in comparison with pure or weathered SMS from *Pleurotus* and *Agaricus* cultures (Roy et al., 2015).

6.8 Fermented products

Other traditional products permitted for use in organic agriculture are fermented household wastes or plant matter. In this group, we can also include biogas-digestate-containing plant residues and animal by-products co-digested with materials of plant origin. Agricultural utilization of digestates is one method of managing these by-products (Dahlin et al., 2017).

Biogas digestates may have a beneficial impact on soil properties due to the presence of macro- and microelements and a high organic matter content (Arthurson, 2009; Möller and Müller, 2012). The digestion process and subsequent processing, such as composting, increase the percentage of nitrogen, phosphorus and potassium in the final product (Macias-Corral et al., 2017). Biogas digestates obtained from plants such as maize affect the mineral uptake by crop plants and have an impact on the composition of soil microbial communities (Hupfauf et al., 2016). Biogas digestates used correctly as fertilizer positively impact the growth of cultivated plants, and yields can be similar to those obtained using traditional mineral fertilizers (Barbosa et al., 2014). On the other hand, utilization of the non-separated or liquid phase of biogas plant digestates in high doses may have toxic effects on soil biota and cultivated plants (Stefaniuk et al., 2015).

6.9 Stillage and stillage extract

Materials obtained during ethanol production are suitable as an effective soil amendment in field and horticultural crops. Stillage is a rich source of residual sugars, organic nitrogen forms and other nutrients whose content is associated with the feedstock material used for fermentation (Wilkie et al., 2000; Alotaibi et al., 2014,).

These properties make stillage and its derivatives a good and promising fertilizer, but utilization of raw stillage is restricted by some legal rules due to its potential negative effect on soil properties (Sajbrt et al., 2010; Fuess and Garcia, 2014). Good results with the use of stillage have been obtained in canola and wheat production (Qian et al., 2011; Alotaibi et al., 2014).

6.9.1 Solid phase residues from biogas production

The solid phase of digestate obtained from biogas production also has beneficial properties. Good-quality feedstock allows the obtaining of a digestate suitable for use in agriculture (Al Seadi et al., 2013). Utilization of the separated solid phase of biogas digestate or stillage is one way of alleviating eco-toxic effects of biogas by-products on plants and soil organisms. Application of the solid phase stimulates root growth, whereas liquid phase digestates show inhibiting effects on the growth of roots in some plants (Stefaniuk et al., 2015). Separated, dried solid phase wastes have a higher C:N ratio in comparison with the liquid or non-separated residues. Also their mineral content is associated with the nature of the biogas digestate.

Residues from biogas production can also be used as feedstock for biochar production, but there is a risk of potential toxicity of biochar for soil organisms associated both with the nature of feedstock material production and pyrolysis temperature. For example, Stefaniuk et al. (2016) showed that biochar obtained from non-separated digestate, pyrolysed at 800°C, is much more toxic for soil organisms in comparison with the separated, solid phase of digestate prepared at 400°C.

6.10 Leonardites

Leonardite is naturally oxidized lignite (brown coal), raw organic sediment rich in humic and fulvic acids, mainly obtained as a by-product of near-surface mining activities. It is used as a raw material as a soil conditioner and for the production of pure humic acid preparations, humates (e.g. potassium or iron salts of humic acids) (Kovács et al., 2013). The production of humic acids is conducted with strong alkali or using physical methods (Canieren et al., 2017).

Humic acids stabilize ion exchange in soils and are very useful in contaminated soils. The effects of humic acids on plant growth and nutrient uptake are associated with physical properties of humic acids, especially their molecular weight (Qian et al., 2015; Sun et al., 2016). Leonardites, or humic acids, accelerate microbial activity in soils, including decomposition of organic pollutants (Tejeda-Agredano et al., 2014). Humic substances extracted from leonardites also have bio-stimulant properties due to the presence of polycyclic hydrocarbons, similar in structure to plant growth regulators (Conselvan et al., 2017).

Similar properties as a raw material are exhibited by brown coal (lignite), which is also a good source of humic substances used as soil improvers. Direct application of lignite to the soil is not recommended due to economic considerations, especially the high cost of long-distance transport, in relation to raw lignite costs and the negative impact of raw lignite on soil properties. Possible negative effects on soil microbial communities or nutrient cycling are visible mainly on a short-time scale. For example, lignite added to a sandy or clay soil decreases microbial activity, measured as respiration, and increases the activity of peroxidase and phenol oxidase enzymes (Kim Thi Tran et al., 2015). A much more effective way is to process lignite into concentrated easy-to-use preparations, rich in humic acids. The methods of extraction of humic acids from brown coal are similar to those used for leonardite, but sometimes with additional treatments for better quality and usability of the final product (Doskočil et al., 2014; Ozkan and Ozkan, 2016). Brown coal and peat, mentioned below, are promising sources of humic acids for the organic fertilizer industry (Liu et al., 2014).

6.11 Peat

Peat is a material of moor origin with use limited to horticulture in organic agriculture in the EU, used as a soil conditioner, a growth substrate for plants, or as soil mulch. Peat is also a source of substances that promote plant growth, for example, humic acids, and phytohormone-like organic substances (Klavins and Purmalis, 2013; Boguta et al., 2016; Szajdak, 2016). Peat is used as a raw material for the production of humic acid preparations and fertilizers for use in agriculture and horticulture (Saito and Seckler, 2014; Agafonova et al., 2015). On the EU market, there are some products obtained from peat, for example, a Latvian liquid preparation called Humate Green OK Universal-Pro and similar products by the same producer, prepared on the basis of peat extracts; it can be applied directly to the soil or on plants to increase the nutritional effect.

6.12 Sapropel and similar sediments

Sapropel refers to sediments rich in organic matter formed under exclusion of oxygen in stagnant water bodies. Apart from humic acids, sapropels contain macro- and microelements, simple organic molecules like amino acids, vitamins, antibiotics and other substances that promote the growth of plants (Blečić et al., 2014; Grantina-levina et al., 2014; Szajdak, 2016). The amounts of organic substances are associated with the conditions of sapropel formation and the degree of its transformation (Dmitrieva et al., 2015). A characteristic feature of some sapropels is the number of microorganisms that mineralize the organic substances to form sapropel (Tretjakova et al., 2015). These properties make sapropels a good substrate for the production of new organic fertilizers (Agafonova et al., 2015). For example, the sapropel-based organic fertilizer 'Humin Plus' is composed of sapropel, peat and chicken manure, additionally treated for improved effectiveness (Ostrovskij, 2014). Due to the high content of nitrogen in some sapropels, there is a need to determine this content and use sapropels correctly to avoid over-fertilization with nitrogen according to the 91/676/EEB rules (Klepeckas and Januškaitienė, 2017). For organic agriculture use in the EU, only fresh water sediments are allowed. Experiments have shown that sapropel is a good material positively affecting soil quality, plant growth and yield quality parameters. Applications of sapropel increase the chlorophyll content in cereals, such as wheat and barley (Klepeckas and Januškaitienė, 2017). Sapropel-based fertilizer 'Humin Plus' increased the yields of tested crops by 23–25% in wheat, 18–27% in corn, 11–32% in sugar beet, 32% in potato and 45–55% in canola, depending on the location of the experiment (Ostrovskij, 2014).

7 Composts

Composts are produced from separated domestic residues, but there is a possibility to use other substrates like mushroom culture wastes, vegetable matter or other materials alone or mixed with farmyard manure. Composting is also a good way to stabilize nutrient content and dehydration of animal manures. Nowadays, compost is one of the organic fertilizers allowing utilization of organic wastes during the composting process (Yu et al., 2016). Yearly compost applications promote plant growth and reduce the risk of soil and groundwater contamination with nitrates (Toselli et al., 2013). Long-term compost application can increase organic carbon content and improve soil structure, mainly due to the formation of soil macro-aggregates, which are fundamental as habitats for soil microorganisms (Zhang et al., 2014). Effects on yields do depend on the nutrient contents, their availability as well as the crop and its environment. Applications of compost can result in the increase of the yield of some crops, for example, peas (*Pisum sativum* L.), but can also have no impact on the yields of other crops, such as oats (*Avena sativa* L.) (Jannoura et al., 2014).

As a slow nutrient releasing agent, applications of compost as a fertilizer or a component of growing media can increase the yield and quality of harvested fruits and vegetables. The use of compost resulted in a larger size and higher number of 'extra class' strawberry fruits of the cultivars 'Senga Sengana', Kent' and 'Elsanta' in comparison with the control (Frąc et al., 2009). Strawberry plants cv. 'Marmolada' and 'Maya' planted in a compost–peat mixture gave higher yields than those planted in peat; cv. 'Patty' gave higher yields in the peat control, whereas in the case of cv. 'Irma', the effect was correlated with the compost content in the medium – with 25% compost, the yields were better in peat, and with 50% compost in the medium, the yield was lower than that in peat (Altieri et al., 2014).

Compost application has a big impact, mainly positive, on soil microbiota such as soil bacteria or fungi, but compost can also decrease the colonization rate of AMF (Derkowska et al., 2008; Zhen et al., 2014).

Vermicompost is the result of conversion of organic residues by earthworms into humus-like substances. There are many types of raw material used for vermicompost production: precomposted animal faeces, crop residues, food processing by-products, industrial residues and others (Sim and Wu, 2010; Das et al., 2016; Domínguez et al., 2017). Vermicompost can enhance soil fertility physically, chemically and biologically. In physical terms, a vermicompost-treated soil has better aeration, porosity, bulk density, drainage and water retention. Chemical properties such as pH, electrical conductivity and organic matter content are also improved for better crop yield (Lim et al., 2015). Nutrient composition in vermicompost depends on the raw material used as feedstock and the species of earthworms used for the vermicomposting process (Lim et al., 2015; Joshi et al., 2015). Vermicompost is a better nutritional source than traditional composts due to its increased rate of mineralization and degree of humification by the action of earthworms. Nutrients are present in forms readily available to plants. The beneficial properties of vermicomposts are increased by the presence of fungi, bacteria and actinomycetes, producers of plant growth regulators and other substances which can affect plant growth and yields (Joshi et al., 2015).

When applied on their own or with selected strains of beneficial microorganisms, vermicomposts have a positive effect on plant growth and yield (Yang et al., 2015b), and can be a part of sustainable, integrated or certified organic production (Bharti et al., 2016a; Beck et al., 2016).

Vermicomposts are also a raw material for the production of liquid formulations, which can have a positive impact on plant growth, but are easier to use for field applications, especially in perennial, for example, pomological crops (Singh et al., 2010; Grzyb et al., 2012; Rozpara et al., 2014; Mosa et al., 2016). Additions of other natural compounds can improve the survivorship of beneficial bacterial strains in vermicompost-based formulations (Kalra et al., 2010).

8 Untreated minerals and by-products of selected industrial processes

Untreated minerals are not strictly organic fertilizers, but are allowed for application in organic agriculture as a source of phosphorus, potassium, magnesium, calcium and microelements.

This category covers crude potassium salt or kainite, potassium sulphate (K_2SO_4) possibly containing Mg salts, calcium carbonate, magnesium and calcium carbonate, magnesium sulphate (kieserite), calcium chloride solution, calcium sulphate (gypsum), industrial lime from sugar production, industrial lime from vacuum salt production, elemental sulphur and trace elements as inorganic micronutrients listed in EC Regulation 2003/2003 (Ciesielska et al., 2011). Untreated minerals are also used as raw materials for the production of organic fertilizers.

9 Biochar

Biochar is a new promising material for organic agriculture, which can be used as a fertilizer, soil improver or pollutant binder, alone or in a mixture with other amendments. Currently, it is not present on the list of products allowed for use in organic agriculture in the EU. Biochar is a material made from organic matter, pyrolysed at high temperature in the absence of oxygen. Charred materials include materials of plant origin, such as wood residues, crop residues, microalgae biomass, animal residues (bones or meals), animal manure and mixed materials, for example, food industry wastes or sewage sludge (Chan et al., 2007, 2008; Sohi et al., 2010; Bird et al., 2011; Enders et al., 2012; Angst et al., 2014; Hosseini Bai et al., 2015). The chemical and physical characteristics of biochars vary depending on the feedstock material and conditions of the conversion, including temperature, time of processing and the technology used (Lehmann et al., 2011; Cantrell et al., 2012; Baronti et al., 2014; Akhter et al., 2015). Biochars produced in different plants but from the same biomass and under similar pyrolysis conditions can result in various properties of the final product (Spokas et al., 2012a). The final product is a material that can have residual (or relic) structures of the original feedstock material or no residual structures (Spokas, 2010). The physical structure of biochars affects the organic and inorganic composition: the pH can range from 5.6 to 13.0, the C content from 33.0% to 82.7%, N content from 0.1% to 6.0% and the C: N ratio from 19 to 221 (Jha et al., 2010; Spokas, 2010; Spokas et al., 2012b). Biochar can also contain appreciable quantities of P, K, Ca, Mg and micronutrients (Cu, Zn, Fe, Mn), with ashes accounting for 5–60% of the weight, depending on the source of the biomass and pyrolysis conditions (Cheng et al., 2008; Enders et al., 2012).

The main goal of biochar applications in previous years was carbon sequestration in soil deposits (Jha et al., 2010). Now the biochar utilization is also focused on increasing crop yields (Jeffery et al., 2011). The main focus nowadays is also to increase soil fertility (Atkinson et al., 2010) using biochar-induced specific properties of soil and biochar's impact on soil microbiota (Steiner et al., 2008, 2009; Anderson et al., 2011; Parvage et al., 2013; Vanek et al., 2016). Biochar can also be used as a component of growing media for plant production, especially as an alternative to non-renewable materials like peat (Kern et al., 2017). Biochars of animal origin are a suitable source of phosphorus and other macro-elements (Siebers et al., 2014).

10 Conclusion

Utilization of organic or non-processed mineral fertilizers is obligatory in organic agriculture. The wide range of products allows correct soil management and plant fertilization, according to sustainable agricultural practice rules. Organic agriculture is also an area with scope for innovative product creation like organic fertilizers enriched with strains of beneficial microorganisms, for example soil bacteria, filamentous or mycorrhizal fungi. Utilization of organic fertilizers will contribute to the development of organic and sustainable nutrient management strategies and cultivation measures in agriculture.

Organic agriculture also allows better utilization of neglected resources like agricultural or food industry wastes, meat industry residues or biogas station residues. On the market there are a number of products based on processed organic wastes, but utilization rate of these wastes is not adequate to the amount of produced wastes.

Innovations from organic agriculture will spread to other areas of human activity, like bio-fortification of conventionally managed arable fields, restoration of degraded soils and industrial areas or in creation of new green spaces in urbanised areas.

11 Where to look for further information

The present and future research in the area of organic fertilizers will be focused on the utilization of a wide range of natural resources and food industry processing wastes for plant fertilization and soil improvement. The group of industry processing wastes as raw materials for fertilizer production is especially valuable, because its utilization will help to avoid air, water and soil contamination from decayed organic residues. The better utilization of wastes within one process should deliver a raw material for further processes. For example, wastes from biogas production should be used as raw material for production of organic soil improvers. Other wastes, like animal bones or fur feather and horn wastes, can serve as raw materials for the production of slow release of phosphorus or nitrogen fertilizers.

There is a need to develop innovative biofertilizers, new fertilization techniques and nutrient management strategies, which will be applicable in modern, environmentally friendly, sustainable agriculture.

12 References

Abouziena, H. F. and Radwan, S. M. (2015), Allelopathic effects of sawdust, rice straw, bur- clover weed and cogongrass on weed control and development of onion, *International Journal of ChemTech Research*, 7, pp. 337–45.

Agafonova, L., Alsina, I., Sokolov, G., Kovrik, S., Bambalov, N., Apse, J. and Rak, M. (2015), New kinds of sapropel and peat based fertilizers, in: *Environment. Technology. Resources.* Rezekne, Latvia, Volume 2, pp. 20–6.

Ahmad, F., Ahmad, I. and Khan, M. S. (2008), Screening of free-living rhizospheric bacteria for their multiple plant growth promoting activities, *Microbiological Research*, 163(2), pp. 173–81. DOI:10.1016/j.micres.2006.04.001.

Akhter, A., Hage-Ahmed, K., Soja, G. and Steinkellner, S. (2015), Compost and biochar alter mycorrhization, tomato root exudation, and development of *Fusarium oxysporum* f. sp. *lycopersici*, *Frontiers in Plant Science*, 6, pp. 529. DOI:10.3389/fpls.2015.00529.

Al Seadi, T., Drosg, B., Fuchs, W., Rutz, D. and Janssen, R. (2013), Biogas digestate quality and utilization, in: Murphy, J. and Baxter, D. (Eds), *The Biogas Handbook: Science, Production and Applications.* Sawston/Cambridge, UK: Woodhead Publishing, pp. 267–301.

Alotaibi, K. D., Schoenau, J. J. and Hao, X. (2014), Fertilizer potential of thin stillage from wheat-based ethanol production, *BioEnergy Research*, 7(4), pp. 1421–9. DOI:10.1007/s12155-014-9473-1.

Altieri, R., Esposito, A., Baruzzi, G. and Nair, T. (2014), Corroboration for the successful application of humified olive mill waste compost in soilless cultivation of strawberry, *International Biodeterioration & Biodegradation*, 88, pp. 118–24. DOI:10.1016/j.ibiod.2013.12.006.

Anderson, C. R., Condron, L. M., Clough, T. J., Fiers, M., Stewart, A., Hill, R. A. and Sherlock, R. R. (2011), Biochar induced soil microbial community change: Implications for biogeochemical cycling of carbon, nitrogen and phosphorus, *Pedobiologia*, 54(5–6), pp. 309–20. DOI:10.1016/j.pedobi.2011.07.005.

Angst, T. E., Six, J., Reay, D. S. and Sohi, S. P. (2014), Impact of pine chip biochar on trace greenhouse gas emissions and soil nutrient dynamics in an annual ryegrass system in California, *Environmental Benefits and Risks of Biochar Application to Soil*, 191, pp. 17–26. DOI:10.1016/j.agee.2014.03.009.

Anusuya, S. and Sathiyabama, M. (2016), Effect of chitosan on growth, yield and curcumin content in turmeric under field condition, *Biocatalysis and Agricultural Biotechnology*, 6, pp. 102–6. DOI:10.1016/j.bcab.2016.03.002.

Arthurson, V. (2009), Closing the global energy and nutrient cycles through application of biogas residue to agricultural land-potential benefits and drawback, *Energies*, 2(2), pp. 226–42.

Arun, K. S. (2007), *Bio-fertilizers for Sustainable Agriculture. Mechanism of P-Solubilization*. Sixth Edition, Jodhpur, India: Agribios Publishers, pp. 196–7.

Atkinson, C. J., Fitzgerald, J. D. and Hipps, N. A. (2010), Potential mechanisms for achieving agricultural benefits from biochar application to temperate soils: A review, *Plant and Soil*, 337(1), pp. 1–18. DOI:10.1007/s11104-010-0464-5.

Aznar, A., Chen, N. W. G., Rigault, M., Riache, N., Joseph, D., Desmaële, D., Mouille, G., Boutet, S., Soubigou-Taconnat, L., Renou, J. P., Thomine, S., Expert, D. and Dellagi, A. (2014), Scavenging iron: A novel mechanism of plant immunity activation by microbial siderophores, *Plant Physiology*, 164(4), pp. 2167–83. DOI:10.1104/pp.113.233585.

Babik, J., Babik, I. and Kaniszewski, S. (2013), New biodegradable agro-fleece for soil mulching in organic vegetable production, in: De Neve, S. (Ed.), *NUTRIHORT: Nutrient Management, Innovative Techniques and Nutrient Legislation in Intensive Horticulture for an Improved Water Quality*. Ghent, Netherlands, pp. 343–9.

Barbosa, D. B. P., Nabel, M. and Jablonowski, N. D. (2014), Biogas-digestate as nutrient source for biomass production of *Sida hermaphrodita*, *Zea mays* L. and *Medicago sativa* L., *Energy Procedia*, 59, pp. 120–6. European Geosciences Union General Assembly 2014, EGU Division Energy, Resources & the Environment (ERE). DOI:10.1016/j.egypro.2014.10.357.

Baronti, S., Vaccari, F. P., Miglietta, F., Calzolari, C., Lugato, E., Orlandini, S., Pini, R., Zulian, C. and Genesio, L. (2014), Impact of biochar application on plant water relations in *Vitis vinifera* (L.), *European Journal of Agronomy*, 53, pp. 38–44. DOI:10.1016/j.eja.2013.11.003.

Basak, A. (2008), Effect of preharvest treatment with seaweed products, Kelpak® and Goëmar BM 86®, on fruit quality in apple, *International Journal of Fruit Science*, 8(1–2), pp. 1–14. DOI:10.1080/15538360802365251.

Beck, J. E., Schroeder-Moreno, M. S., Fernandez, G. E., Grossman, J. M. and Creamer, N. G. (2016), Effects of cover crops, compost, and vermicompost on strawberry yields and nitrogen availability in North Carolina, *HortTechnology*, 26(5), pp. 604–13. DOI:10.21273/HORTTECH03447-16.

Bedini, S., Turrini, A., Rigo, C., Argese, E. and Giovannetti, M. (2010), Molecular characterization and glomalin production of arbuscular mycorrhizal fungi colonizing a heavy metal polluted ash disposal island, downtown Venice, *Soil Biology and Biochemistry*, 42(5), pp. 758–65. DOI:10.1016/j.soilbio.2010.01.010.

Bharti, N., Barnawal, D., Shukla, S., Tewari, S. K., Katiyar, R. S. and Kalra, A. (2016a), Integrated application of *Exiguobacterium oxidotolerans*, *Glomus fasciculatum*, and vermicompost improves growth, yield and quality of *Mentha arvensis* in salt-stressed soils, *Industrial Crops and Products*, 83, pp. 717–28. DOI:10.1016/j.indcrop.2015.12.021.

Bharti, N., Barnawal, D., Wasnik, K., Tewari, S. K. and Kalra, A. (2016b), Co-inoculation of *Dietzia natronolimnaea* and *Glomus intraradices* with vermicompost positively influences *Ocimum basilicum* growth and resident microbial community structure in salt affected low fertility soils, *Applied Soil Ecology*, 100, pp. 211–25. DOI:10.1016/j.apsoil.2016.01.003.

Bird, M. I., Wurster, C. M., de Paula Silva, P. H., Bass, A. M. and de Nys, R. (2011), Algal biochar – production and properties, *Bioresource Technology*, 102(2), pp. 1886–91. DOI:10.1016/j.biortech.2010.07.106.

Blečić, A., Railić, B., Dubljević, R., Mitrović, D. and Spalevic, V. (2014), Application of sapropel in agricultural production, *Agriculture and Forestry*, 60(2), pp. 243–50.

Boguta, P., D'Orazio, V., Sokołowska, Z. and Senesi, N. (2016), Effects of selected chemical and physicochemical properties of humic acids from peat soils on their interaction mechanisms with copper ions at various pHs, *Journal of Geochemical Exploration*, 168, pp. 119–26. DOI:10.1016/j.gexplo.2016.06.004.

Bona, E., Lingua, G., Manassero, P., Cantamessa, S., Marsano, F., Todeschini, V., Copetta, A., D'Agostino, G., Massa, N., Avidano, L., Gamalero, E. and Berta, G. (2015), AM fungi and PGP pseudomonads increase flowering, fruit production, and vitamin content in strawberry grown at low nitrogen and phosphorus levels, *Mycorrhiza*, 25(3), pp. 181–93. DOI:10.1007/s00572-014-0599-y.

Bona, E., Cantamessa, S., Massa, N., Manassero, P., Marsano, F., Copetta, A., Lingua, G., D'Agostino, G., Gamalero, E. and Berta, G. (2017), Arbuscular mycorrhizal fungi and plant growth-promoting pseudomonads improve yield, quality and nutritional value of tomato: A field study, *Mycorrhiza*, 27(1), pp. 1–11. DOI:10.1007/s00572-016-0727-y.

Born, Y., Remus-Emsermann, M. N., Bieri, M., Kamber, T., Piel, J. and Pelludat, C. (2016), Fe^{2+} chelator proferrorosamine A: A gene cluster of Erwinia rhapontici P45 involved in its synthesis and its impact on growth of Erwinia amylovora CFBP1430, *Microbiology*, 162(2), pp. 236–45.

Boursier, H., Béline, F. and Paul, E. (2005), Piggery wastewater characterisation for biological nitrogen removal process design, *Bioresource Technology*, 96(3), pp. 351–8. DOI:10.1016/j.biortech.2004.03.007.

Briceño-Domínguez, D., Hernández-Carmona, G., Moyo, M., Stirk, W. and van Staden, J. (2014), Plant growth promoting activity of seaweed liquid extracts produced from Macrocystis pyrifera under different pH and temperature conditions, *Journal of Applied Phycology*, 26(5), pp. 2203–10. DOI:10.1007/s10811-014-0237-2.

Canieren, O., Karaguzel, C. and Aydin, A. (2017), Effect of physical pre-enrichment on humic substance recovery from leonardite, *Physicochemical Problems of Mineral Processing*, 53(1), pp. 502–14.

Cantrell, K. B., Hunt, P. G., Uchimiya, M., Novak, J. M. and Ro, K. S. (2012), Impact of pyrolysis temperature and manure source on physicochemical characteristics of biochar, *Bioresource Technology*, 107, pp. 419–28. DOI:10.1016/j.biortech.2011.11.084.

Cassán, F., Vanderleyden, J. and Spaepen, S. (2014), Physiological and agronomical aspects of phytohormone production by model Plant-Growth-Promoting Rhizobacteria (PGPR) belonging to the genus Azospirillum, *Journal of Plant Growth Regulation*, 33(2), pp. 440–59. DOI:10.1007/s00344-013-9362-4.

Cavagnaro, T. R., Bender, S. F., Asghari, H. R. and Heijden, M. G. A. van der (2015), The role of arbuscular mycorrhizas in reducing soil nutrient loss, *Trends in Plant Science*, 20(5), pp. 283-90. DOI:10.1016/j.tplants.2015.03.004.

Chan, K. Y., Van Zwieten, L., Meszaros, I., Downie, A. and Joseph, S. (2007), Agronomic values of greenwaste biochar as a soil amendment, *Soil Research*, 45(8), pp. 629-34.

Chan, K. Y., Van Zwieten, L., Meszaros, I., Downie, A. and Joseph, S. (2008), Using poultry litter biochars as soil amendments, *Soil Research*, 46(5), pp. 437-44.

Chantal, K., Shao, X., Jing, B., Yuan, Y., Hou, M. and Liao, L. (2013), Effects of effective microorganisms (EM) and bio-organic fertilizers on growth parameters and yield quality of flue-cured tobacco (Nicotiana tabacum), *Journal of Food, Agriculture and Environment*, 11(2), pp. 1212-15.

Chen, Y. P., Rekha, P. D., Arun, A. B., Shen, F. T., Lai, W.-A. and Young, C. C. (2006), Phosphate solubilizing bacteria from subtropical soil and their tricalcium phosphate solubilizing abilities, *Applied Soil Ecology*, 34(1), pp. 33–41. DOI:10.1016/j.apsoil.2005.12.002.

Cheng, C.-H., Lehmann, J., Thies, J. E. and Burton, S. D. (2008), Stability of black carbon in soils across a climatic gradient, *Journal of Geophysical Research: Biogeosciences*, 113(G2), pp. G02027. DOI:10.1029/2007JG000642.

Ciesielska, J., Malusa, E. and Sas-Paszt, L. (2011), Środki ochrony roślin stosowane w rolnictwie ekologicznym, *Komentarz Do Załącznika II Rozporządzenia Komisji (WE) Nr*, 889(2008), pp. 17-19.

Chetverikov, S. P. and Loginov, O. N. (2009), New metabolites of Azotobacter vinelandii exhibiting antifungal activity, *Microbiology*, 78(4), pp. 428-32. DOI:10.1134/S0026261709040055.

Colla, G., Rouphael, Y., Canaguier, R., Svecova, E. and Cardarelli, M. (2014), Biostimulant action of a plant-derived protein hydrolysate produced through enzymatic hydrolysis, *Frontiers in Plant Science*, 5, pp. 448. DOI:10.3389/fpls.2014.00448.

Colla, G., Nardi, S., Cardarelli, M., Ertani, A., Lucini, L., Canaguier, R. and Rouphael, Y. (2015), Protein hydrolysates as biostimulants in horticulture, *Biostimulants in Horticulture*, 196, pp. 28-38. DOI:10.1016/j.scienta.2015.08.037.

Conselvan, G. B., Pizzeghello, D., Francioso, O., Di Foggia, M., Nardi, S. and Carletti, P. (2017), Biostimulant activity of humic substances extracted from leonardites, *Plant Soil*, 420(1-2), pp. 119-34. DOI:10.1007/s11104-017-3373-z.

Corte, L., Dell'Abate, M. T., Magini, A., Migliore, M., Felici, B., Roscini, L., Sardella, R., Tancini, B., Emiliani, C., Cardinali, G. and Benedetti, A. (2014), Assessment of safety and efficiency of nitrogen organic fertilizers from animal-based protein hydrolysates–a laboratory multidisciplinary approach, *Journal of the Science of Food and Agriculture*, 94(2), pp. 235-45. DOI:10.1002/jsfa.6239.

Cretoiu, M. S., Korthals, G. W., Visser, J. H. M. and van Elsas, J. D. (2013), Chitin amendment increases soil suppressiveness toward plant pathogens and modulates the actinobacterial and oxalobacteraceal communities in an experimental agricultural field, *Applied and Environmental Microbiology*, 79(17), pp. 5291-301. DOI:10.1128/AEM.01361-13.

Crouch, I. J. and van Staden, J. (1993), Evidence for the presence of plant growth regulators in commercial seaweed products, *Plant Growth Regulation*, 13(1), pp. 21–9. DOI:10.1007/BF00207588.

Crouch, I. J., Beckett, R. P. and Staden, J. (1990), Effect of seaweed concentrate on the growth and mineral nutrition of nutrient-stressed lettuce, *Journal of Applied Phycology*, 2(3), pp. 269–72. DOI:10.1007/BF02179784.

Cuijpers, W. J. M. and Hospers-Brands, A. (2008), *Hulpmeststoffen: beschikbaarheid en opname van stikstof in de biologische teelt van zomertarwe*. Driebergen, the Netherlands: Louis Bolk Instituut, 33pp.

Dahlin, J., Nelles, M. and Herbes, C. (2017), Biogas digestate management: Evaluating the attitudes and perceptions of German gardeners towards digestate-based soil amendments, *Resources, Conservation and Recycling*, 118, pp. 27–38. DOI:10.1016/j. resconrec.2016.11.020.

Das, N. and Mukherjee, M. (2007), Cultivation of Pleurotus ostreatus on weed plants, *Bioresource Technology*, 98(14), pp. 2723–6. DOI:10.1016/j.biortech.2006.09.061.

Das, V., Satyanarayan, S. and Satyanarayan, S. (2016), Value added product recovery from sludge generated during gum arabic refining process by vermicomposting, *Environmental Monitoring and Assessment*, 188(9), pp. 523. DOI:10.1007/ s10661-016-5516-8.

Davison, J., Moora, M., Öpik, M., Adholeya, A., Ainsaar, L., Bâ, A., Burla, S., Diedhiou, A. G., Hiiesalu, I., Jairus, T., Johnson, N. C., Kane, A., Koorem, K., Kochar, M., Ndiaye, C., Pärtel, M., Reier, Ü., Saks, Ü., Singh, R., Vasar, M. and Zobel, M. (2015), Global assessment of arbuscular mycorrhizal fungus diversity reveals very low endemism, *Science*, 349(6251), pp. 970. DOI:10.1126/science.aab1161.

Depardieu, C., Prémont, V., Boily, C. and Caron, J. (2016), Sawdust and bark-based substrates for soilless strawberry production: Irrigation and electrical conductivity management, *PLoS ONE*, 11(4), pp. e0154104 (1–20). DOI:10.1371/journal. pone.0154104.

Derkowska, E., Sas-Paszt, L., Sumorok, B., Szwonek, E. and Gluszek, S. (2008), The influence of mycorrhization and organic mulches on mycorrhizal frequency in apple and strawberry roots, *Journal of Fruit and Ornamental Plant Research*, 16, pp. 227–42.

Derkowska, E., Sas Paszt, L. S., Harbuzov, A. and Sumorok, B. (2015a), Root growth, mycorrhizal frequency and soil microorganisms in strawberry as affected by biopreparations, *Advances in Microbiology*, 5(1), pp. 65–73.

Derkowska, E., Sas Paszt, L. and Szwonek, E. (2015b), Influence of biological products on the growth and development of large-fruited cranberry under greenhouse conditions, *Folia Horticulturae*, 27(1), pp. 71–7. DOI:10.1515/fhort-2015-0016.

Dhar, D. W., Prasanna, R., Pabbi, S. and Vishwakarma, R. (2015), Significance of cyanobacteria as inoculants in agriculture, in: Das, D. (Ed.), *Algal Biorefinery: An Integrated Approach*. Cham, Switzerland: Springer International Publishing, pp. 339–74.

Di Stasio, E., Maggio, A., Ventorino, V., Pepe, O., Raimondi, G. and De Pascale, S. (2017), Free-living (N2)-fixing bacteria as potential enhancers of tomato growth under salt stress. *Acta Horticulturae*, 1164, pp. 151–6. DOI:10.17660/ActaHortic.2017.1164.19

Dmitrieva, E., Efimova, E., Siundiukova, K. and Perelomov, L. (2015), Surface properties of humic acids from peat and sapropel of increasing transformation, *Environmental Chemistry Letters*, 13(2), pp. 197–202. DOI:10.1007/s10311-015-0497-3.

Domínguez, J., Sanchez-Hernandez, J. C. and Lores, M. (2017), Vermicomposting of winemaking by-products, in: Galanakis, C. M. (Ed.), *Handbook of Grape Processing By-Products*. London, UK: Academic Press, Elsevier, pp. 55–78.

Doskočil, L., Grasset, L., Válková, D. and Pekař, M. (2014), Hydrogen peroxide oxidation of humic acids and lignite, *Fuel*, 134, pp. 406–13. DOI:10.1016/j.fuel.2014.06.011.

Duong, D. H., Ngo, X. Q., Do, D. G., Le, T. A. H., Nguyen, V. T. and Nic, S. (2014), Effective control of neem (*Azadirachta indica* A. Juss) cake to plant parasitic nematodes and fungi in black pepper diseases in vitro, *Journal of Vietnamese Environment*, 6 (3), pp. 233–8.

Ebrahiminezhad, A., Manafi, Z., Berenjian, A., Kianpour, S. and Ghasemi, Y. (2017), Iron-reducing bacteria and iron nanostructures, *Journal of Advanced Medical Sciences and Applied Technologies*, 3(1), pp. 9–16.

Egamberdiyeva, D. (2007), The effect of plant growth promoting bacteria on growth and nutrient uptake of maize in two different soils, *Applied Soil Ecology*, 36(2–3), pp. 184–9. DOI:10.1016/j.apsoil.2007.02.005.

Emerson, J. K. and Roark, A. M. (2007), Composition of guano produced by frugivorous, sanguivorous, and insectivorous bats, *Acta Chiropterologica*, 9(1), pp. 261–7. DOI:10.3161/1733-5329(2007)9[261:COGPBF]2.0.CO;2.

Enders, A., Hanley, K., Whitman, T., Joseph, S. and Lehmann, J. (2012), Characterization of biochars to evaluate recalcitrance and agronomic performance, *Bioresource Technology*, 114, pp. 644–53. DOI:10.1016/j.biortech.2012.03.022.

Fan, D., Hodges, D. M., Zhang, J., Kirby, C. W., Ji, X., Locke, S. J., Critchley, A. T. and Prithiviraj, B. (2011), Commercial extract of the brown seaweed *Ascophyllum nodosum* enhances phenolic antioxidant content of spinach (*Spinacia oleracea* L.) which protects *Caenorhabditis elegans* against oxidative and thermal stress, *Food Chemistry*, 124(1), pp. 195–202. DOI:10.1016/j.foodchem.2010.06.008.

Fernández-Delgado Juárez, M., Gómez-Brandón, M. and Insam, H. (2015), Merging two waste streams, wood ash and biowaste, results in improved composting process and end products, *Science of the Total Environment*, 511, pp. 91–100. DOI:10.1016/j.scitotenv.2014.12.037.

Fox, A. R., Soto, G., Valverde, C., Russo, D., Lagares, A., Zorreguieta, Á., Alleva, K., Pascuan, C., Frare, R., Mercado-Blanco, J., Dixon, R. and Ayub, N. D. (2016), Major cereal crops benefit from biological nitrogen fixation when inoculated with the nitrogen-fixing bacterium Pseudomonas protegens Pf- 5 X940, *Environmental Microbiology*, 18(10), pp. 3522–34. DOI:10.1111/1462-2920.13376.

Frąc, M., Michalski, P. and Sas Paszt, L. (2009), The effect of mulch and mycorrhiza on fruit yield and size of three strawberry cultivars, *Journal of Fruit and Ornamental Plant Research*, 17(2), pp. 85–93.

Fuess, L. T. and Garcia, M. L. (2014), Implications of stillage land disposal: A critical review on the impacts of fertigation, *Journal of Environmental Management*, 145, pp. 210–29. DOI:10.1016/j.jenvman.2014.07.003.

Ghani, W. A. W. A. K., Mohd, A., da Silva, G., Bachmann, R. T., Taufiq-Yap, Y. H., Rashid, U. and Al-Muhtaseb, A. H. (2013), Biochar production from waste rubber-wood-sawdust and its potential use in C sequestration: Chemical and physical characterization, *Industrial Crops and Products*, 44, pp. 18–24. DOI:10.1016/j.indcrop.2012.10.017.

Ghasemi, S., Khoshgoftarmanesh, A., Hadadzadeh, H. and Afyuni, M. (2013), Synthesis, characterization, and theoretical and experimental investigations of zinc(II)–amino acid complexes as ecofriendly plant growth promoters and highly bioavailable sources of zinc, *Journal of Plant Growth Regulation*, 32(2), pp. 315–23. DOI:10.1007/s00344-012-9300-x.

Glinicki, R., Sas-Paszt, L. and Jadczuk-Tobjasz, E. (2011), The effect of microbial inoculation with EM-Farming inoculum on the vegetative growth of three strawberry cultivars, *Annals of Warsaw University of Life Sciences- SGGW. Horticulture and Landscape Architecture*, 32, pp. 3–14.

Grantina-Ievina, L., Karlsons, A., Andersone-Ozola, U. and Ievinsh, G. (2014), Effect of freshwater sapropel on plants in respect to its growth-affecting activity and cultivable microorganism content, *Zemdirbyste-Agriculture*, 101(4), pp. 355–66.

Grzyb, Z. S., Piotrowski, W., Bielicki, P., Sas Paszt, L. and Malusà, E. (2012), Effect of different fertilizers and amendments on the growth of apple and sour cherry rootstocks in an organic nursery, *Journal of Fruit and Ornamental Plant Research*, 20(1), pp. 43–53. DOI:10.2478/v10290-012-0004-x.

Grzyb, Z. S., Sas Paszt, L., Piotrowski, W. and Malusa, E. (2015), The influence of mycorrhizal fungi on the growth of apple and sour cherry maidens fertilized with different bioproducts in the organic nursery, *Journal of Life Sciences*, 9, pp. 221–8.

Gusain, Y. S., Kamal, R., Mehta, C. M., Singh, U. S. and Sharma, A. K. (2015), Phosphate solubilizing and indole- 3-acetic acid producing bacteria from the soil of Garhwal Himalaya aimed to improve the growth of rice, *Journal of Environmental Biology*, 36(1), pp. 301.

Hankins, S. D. and Hockey, H. P. (1990), The effect of a liquid seaweed extract from *Ascophyllum nodosum* (Fucales, Phaeophyta) on the two-spotted red spider mite *Tetranychus urticae*, *Hydrobiologia*, 204–205(1), pp. 555–9. DOI:10.1007/BF00040286.

Haynes, R. J. and Swift, R. S. (1986), Effect of soil amendments and sawdust mulching on growth, yield and leaf nutrient content of highbush blueberry plants, *Scientia Horticulturae*, 29(3), pp. 229–38. DOI:10.1016/0304-4238(86)90066-X.

Hernández-Herrera, R. M., Santacruz-Ruvalcaba, F., Ruiz-López, M. A., Norrie, J. and Hernández-Carmona, G. (2014), Effect of liquid seaweed extracts on growth of tomato seedlings (*Solanum lycopersicum* L.), *Journal of Applied Phycology*, 26(1), pp. 619–28. DOI:10.1007/s10811-013-0078-4.

Hoagland, L., Carpenter-Boggs, L., Granatstein, D., Mazzola, M., Smith, J., Peryea, F. and Reganold, J. P. (2008), Orchard floor management effects on nitrogen fertility and soil biological activity in a newly established organic apple orchard, *Biology and Fertility of Soils*, 45(1), pp. 11. DOI:10.1007/s00374-008-0304-4.

Hosseini Bai, S., Xu, C.-Y., Xu, Z., Blumfield, T., Zhao, H., Wallace, H., Reverchon, F. and Van Zwieten, L. (2015), Soil and foliar nutrient and nitrogen isotope composition (δ15N) at 5 years after poultry litter and green waste biochar amendment in a macadamia orchard, *Environmental Science and Pollution Research*, 22(5), pp. 3803–9. DOI:10.1007/s11356-014-3649-2.

Hu, C. and Qi, Y. (2013), Long-term effective microorganisms application promote growth and increase yields and nutrition of wheat in China, *European Journal of Agronomy*, 46, pp. 63–7. DOI:10.1016/j.eja.2012.12.003.

Hupfauf, S., Bachmann, S., Fernández-Delgado Juárez, M., Insam, H. and Eichler-Löbermann, B. (2016), Biogas digestates affect crop P uptake and soil microbial community composition, 542, pp. 1144–54. Special Issue on Sustainable Phosphorus

Taking Stock: Phosphorus Supply from Natural and Anthropogenic Pools in the 21st Century. DOI:10.1016/j.scitotenv.2015.09.025.

Ibrahim, E. A. and Ramadan, W. A. (2015), Effect of zinc foliar spray alone and combined with humic acid or/and chitosan on growth, nutrient elements content and yield of dry bean (*Phaseolus vulgaris* L.) plants sown at different dates, *Scientia Horticulturae*, 184, pp. 101-5. DOI:10.1016/j.scienta.2014.11.010.

Jakienė, E. (2013), The effect of the microelement fertilizers and biological preparation Terra Sorb Foliar on spring rape crop, *Žemės Ūkio Mokslai*, 20(2), pp. 75-83.

Jannoura, R., Joergensen, R. G. and Bruns, C. (2014), Organic fertilizer effects on growth, crop yield, and soil microbial biomass indices in sole and intercropped peas and oats under organic farming conditions, *European Journal of Agronomy*, 52(Part B), pp. 259-70. DOI:10.1016/j.eja.2013.09.001.

Jayaraj, J., Wan, A., Rahman, M. and Punja, Z. K. (2008), Seaweed extract reduces foliar fungal diseases on carrot, *Crop Protection*, 27 (10), pp. 1360-6. DOI:10.1016/j.cropro.2008.05.005.

Jayaraman, J., Norrie, J. and Punja, Z. (2011), Commercial extract from the brown seaweed *Ascophyllum nodosum* reduces fungal diseases in greenhouse cucumber, *Journal of Applied Phycology*, 23(3), pp. 353-61. DOI:10.1007/s10811-010-9547-1.

Jeffery, S., Verheijen, F. G. A., van der Velde, M. and Bastos, A. C. (2011), A quantitative review of the effects of biochar application to soils on crop productivity using meta-analysis, *Agriculture, Ecosystems & Environment*, 144(1), pp. 175-87. DOI:10.1016/j.agee.2011.08.015.

Jha, P., Biswas, A. K., Lakaria, B. L. and Rao, A. S. (2010), Biochar in agriculture-prospects and related implications, *Current Science*, 99(9), pp. 1218-25.

Ji, R., Dong, G., Shi, W. and Min, J. (2017), Effects of liquid organic fertilizers on plant growth and rhizosphere soil characteristics of Chrysanthemum, *Sustainability*, 9(5), pp. 841 (1-16). DOI:10.3390/su9050841.

Jones, D. L. and Quilliam, R. S. (2014), Metal contaminated biochar and wood ash negatively affect plant growth and soil quality after land application, *Journal of Hazardous Materials*, 276, pp. 362-70. DOI:10.1016/j.jhazmat.2014.05.053.

Jones, S. B., Robbins, C. W. and Hansen, C. L. (1993), Sodic soil reclamation using cottage cheese (acid) whey, *Arid Soil Research and Rehabilitation*, 7(1), pp. 51-61. DOI:10.1080/15324989309381334.

Joshi, R., Singh, J. and Vig, A. (2015), Vermicompost as an effective organic fertilizer and biocontrol agent: Effect on growth, yield and quality of plants, *Reviews in Environmental Science and Bio/Technology*, 14(1), pp. 137-59. DOI:10.1007/s11157-014-9347-1.

Jung, J. Y., Kim, J. S., Ha, S. Y., Choi, J. H. and Yang, J.-K. (2015), Suitability of thermal treated sawdust as replacements for peat moss in horticultural media, *Journal of Agriculture and Life Science*, 49(4), pp. 105-15.

Kalra, A., Chandra, M., Awasthi, A., Singh, A. K. and Khanuja, S. P. S. (2010), Natural compounds enhancing growth and survival of rhizobial inoculants in vermicompost-based formulations, *Biology and Fertility of Soils*, 46(5), pp. 521-4. DOI:10.1007/s00374-010-0443-2.

Kaniszewski, S., Babik, I. and Babik, J. (2013), Pelletized legume plants as fertilizer for vegetables in organic farming, in: De Neve, S. (Ed.), *NUTRIHORT: Nutrient*

Management, Innovative Techniques and Nutrient Legislation in Intensive Horticulture for an Improved Water Quality. Ghent, Netherlans, pp. 330–42.

Kern, J., Tammeorg, P., Shanskiy, M., Sakrabani, R., Knicker, H., Kammann, C., Tuhkanen, E.-M., Smidt, G., Prasad, M., Tiilikkala, K., Sohi, S., Gascó, G., Steiner, C. and Glaser, B. (2017), Synergistic use of peat and charred material in growing media – an option to reduce the pressure on peatlands? *Journal of Environmental Engineering and Landscape Management*, 25(2), pp. 160–74. DOI:10.3846/16486897.2017.1284665.

Khan, W., Rayirath, U., Subramanian, S., Jithesh, M., Rayorath, P., Hodges, D., Critchley, A. T., Craigie, J. S., Norrie, J. and Prithiviraj, B. (2009), Seaweed extracts as biostimulants of plant growth and development, *Journal of Plant Growth Regulation*, 28 (4), pp. 386–99.

Khan, N., Clark, I., Sánchez-Monedero, M. A., Shea, S., Meier, S. and Bolan, N. (2014), Maturity indices in co- composting of chicken manure and sawdust with biochar, *Bioresource Technology*, 168, pp. 245–51. DOI:10.1016/j.biortech.2014.02.123.

Kiers, E. T., Rousseau, R. A., West, S. A. and Denison, R. F. (2003), Host sanctions and the legume-rhizobium mutualism, *Nature*, 425(6953), pp. 78–81. DOI:10.1038/nature01931.

Kim Thi Tran, C., Rose, M. T., Cavagnaro, T. R. and Patti, A. F. (2015), Lignite amendment has limited impacts on soil microbial communities and mineral nitrogen availability, *Applied Soil Ecology*, 95, pp. 140–50. DOI:10.1016/j.apsoil.2015.06.020.

Klavins, M. and Purmalis, O. (2013), Properties and structure of raised bog peat humic acids, *Journal of Molecular Structure*, 1050, pp. 103–13. DOI:10.1016/j.molstruc.2013.07.021.

Kleiber, T., Starzyk, J., Górski, R., Sobieralski, K., Siwulski, M., Rempulska, A. and Sobiak, A. (2014), The studies on applying of Effective Microorganisms (EM) and CRF on nutrient contents in leaves and yielding of tomato, *Acta Scientiarum Polonorum – Hortorum Cultus*, 13 (1), pp. 79–90.

Klepeckas, M. and Januškaitienė, I. (2017), Changes in Triticum aestivum and Hordeum vulgare chlorophyll content and fluorescence parameters under impact of various sapropel concentrations, *Biologija*, 62 (4), pp. 216–26.

Köpke, U., Krämer, J. and Leifert, C. (2007), Pre-harvest strategies to ensure the microbiological safety of fruit and vegetables from manure-based production systems, in: Cooper, J., Niggli, U. and Leifert, C. (Eds), *Handbook of Organic Food Safety and Quality*. Cambridge, UK: Woodhead Publishing Ltd., pp. 413–29.

Koutrotsios, G., Mountzouris, K. C., Chatzipavlidis, I. and Zervakis, G. I. (2014), Bioconversion of lignocellulosic residues by *Agrocybe cylindracea* and *Pleurotus ostreatus* mushroom fungi – assessment of their effect on the final product and spent substrate properties, *Food Chemistry*, 161, pp. 127–35. DOI:10.1016/j.foodchem.2014.03.121.

Kovács, K., Czech, V., Fodor, F., Solti, A., Lucena, J. J., Santos-Rosell, S. and Hernández-Apaolaza, L. (2013), Characterization of Fe–Leonardite complexes as novel natural iron fertilizers, *Journal of Agricultural and Food Chemistry*, 61(50), pp. 12200–10. DOI:10.1021/jf404455y.

Kumar, G. and Sahoo, D. (2011), Effect of seaweed liquid extract on growth and yield of Triticum aestivum var. Pusa Gold, *Journal of Applied Phycology*, 23(2), pp. 251–5. DOI:10.1007/s10811-011-9660-9.

Kumar, V., Kumar, S., Jha, S. K. and Jijeesh, C. M. (2014), Influence of de-oiled seed cakes on seedling performance of East Indian Rosewood (Dalbergia latifoila Roxb.), *Soil Environment*, 33(2), pp. 169-74.

Kumar, V., Menon, S., Agarwal, H. and Gopalakrishnan, D. (2017), Characterization and optimization of bacterium isolated from soil samples for the production of siderophores, *Resource-Efficient Technologies*, 3(4), 434-9. DOI:10.1016/j.reffit.2017.04.004.

Kuwada, K., Kuramoto, M., Utamura, M., Matsushita, I. and Ishii, T. (2006), Isolation and structural elucidation of a growth stimulant for arbuscular mycorrhizal fungus from Laminaria japonica Areschoug, *Journal of Applied Phycology*, 18(6), pp. 795-800.

Laghari, M., Hu, Z., Mirjat, M. S., Xiao, B., Tagar, A. A. and Hu, M. (2016), Fast pyrolysis biochar from sawdust improves the quality of desert soils and enhances plant growth, *Journal of the Science of Food and Agriculture*, 96(1), pp. 199-206. DOI:10.1002/jsfa.7082.

Lehmann, J., Rillig, M. C., Thies, J., Masiello, C. A., Hockaday, W. C. and Crowley, D. (2011), Biochar effects on soil biota - A review, *Soil Biology & Biochemistry*, 43(9), pp. 1812-36. 19th International Symposium on Environmental Biogeochemistry. DOI:10.1016/j.soilbio.2011.04.022.

Lim, S. L., Wu, T. Y., Lim, P. N. and Shak, K. P. Y. (2015), The use of vermicompost in organic farming: Overview, effects on soil and economics, *Journal of the Science of Food and Agriculture*, 95(6), pp. 1143-56. DOI:10.1002/jsfa.6849.

Lima, J. D., Zanetti, S., Nomura, E. S., Fuzitani, E. J., Rozane, D. E. and Iori, P. (2016), Growth and yield of anthurium in response to sawdust mulching, *Ciência Rural*, 46, pp. 440-6.

Lisiecka, J., Knaflewski, M., Spizewski, T., Fraszczak, B., Kaluzewicz, A. and Krzesinski, W. (2011), The effect of animal protein hydrolysate on quantity and quality of strawberry daughter plants cv.'Elsanta', *Acta Scientiarum Polonorum, Hortorum Cultus*, 10, pp. 31-40.

Liu, M., Wang, T., Cheng, Y., Fu, Y., Zhang, P., Xie, M., Sun, M. and Yang, D. (2014), Peat and brown coal resources in China and its potential for developing potassium humate fertilizer, *Earth Science Frontiers*, 5, pp. 25.

Løes, A.-K., Bünemann, E. K., Cooper, J., Hörtenhuber, S., Magid, J., Oberson, A. and Möller, K. (2016), Nutrient supply to organic agriculture as governed by EU regulations and standards in six European countries, *Organic Agriculture*, 7, pp. 395-418. DOI:10.1007/s13165-016-0165-3.

Macias-Corral, M. A., Samani, Z. A., Hanson, A. T. and Funk, P. A. (2017), Co-digestion of agricultural and municipal waste to produce energy and soil amendment, *Waste Management & Research*, 35(9), pp. 991-6. DOI:10.1177/0734242X17715097.

Mahdavi, B. and Rahimi, A. (2013), Seed priming with chitosan improves the germination and growth performance of ajowan (*Carum copticum*) under salt stress, *EurAsian Journal of BioSciences*, 7, 69-76.

Malusà, E., Sas Paszt, L., Głuszek, S. and Ciesielska, J. (2014), Organic fertilizers to sustain soil fertility, in: Sinha, S., Pant, K., Bajpai, S., and Govil, J. N. (Eds), *Fertilizers Technology Vol. I: Synthesis*. Houston, TX: Studium Press LLC, pp. 256-81.

Marinou, E., Chrysargyris, A. and Tzortzakis, N. (2013), Use of sawdust, coco soil and pumice in hydroponically grown strawberry, *Plant, Soil and Environment*, 59(10), pp. 452-9.

Marwaha, S. S. and Kennedy, J. F. (1988), Whey–pollution problem and potential utilization, *International Journal of Food Science and Technology*, 23(4), pp. 323–36. DOI:10.1111/j.1365-2621.1988.tb00586.x.

Masny, A., Basak, A. and Żurawicz, E. (2004), Effects of foliar applications of Kelpak SL and Goëmar BM 86®; preparations on yield and fruit quality in two strawberry cultivars, *Journal of Fruit and Ornamental Plant Research*, 12, pp. 23–7.

Mazzoncini, M., Antichi, D., Tavarini, S., Silvestri, N., Lazzeri, L. and D'Avino, L. (2015), Effect of defatted oilseed meals applied as organic fertilizers on vegetable crop production and environmental impact, *Industrial Crops and Products*, 75(Part A), pp. 54–64. DOI:10.1016/j.indcrop.2015.04.061.

Mobini, M., Khoshgoftarmanesh, A. H. and Ghasemi, S. (2014), The effect of partial replacement of nitrate with arginine, histidine, and a mixture of amino acids extracted from blood powder on yield and nitrate accumulation in onion bulb, *Scientia Horticulturae*, 176, pp. 232–7. DOI:10.1016/j.scienta.2014.07.014.

Möller, K. (2015), Assessment of alternative phosphorus fertilizers for organic farming: Meat and bone meal, Improve-P Factsheet, Universität Hohenheim, ETH Zürich, FiBL, Bioforsk, Universität für Bodenkultur Wien, Newcastle University, University of Copenhagen, pp. 1–8.

Möller, K. and Müller, T. (2012), Effects of anaerobic digestion on digestate nutrient availability and crop growth: A review, *Engineering in Life Sciences*, 12(3), pp. 242–57. DOI:10.1002/elsc.201100085.

Mosa, W.-G., Paszt, L. S., Frąc, M., Trzciński, P., Przybył, M., Treder, W. and Klamkowski, K. (2016), The influence of biofertilization on the growth, yield and fruit quality of cv. Topaz apple trees, *Horticultural Science*, 43(3), pp. 105–11.

Moutia, J.-F. Y., Saumtally, S., Spaepen, S. and Vanderleyden, J. (2010), Plant growth promotion by *Azospirillum* sp. in sugarcane is influenced by genotype and drought stress, *Plant and Soil*, 337(1), pp. 233–42. DOI:10.1007/s11104-010-0519-7.

Nagaraja, H., Chennappa, G., Rakesh, S., Naik, M. K., Amaresh, Y. S. and Sreenivasa, M. Y. (2016), Antifungal activity of *Azotobacter nigricans* against trichothecene-producing *Fusarium* species associated with cereals, *Food Science and Biotechnology*, 25(4), pp. 1197–204. DOI:10.1007/s10068-016-0190-8.

Ngala, B. M., Valdes, Y., dos Santos, G., Perry, R. N. and Wesemael, W. M. L. (2016), Seaweed-based products from *Ecklonia maxima* and *Ascophyllum nodosum* as control agents for the root-knot nematodes *Meloidogyne chitwoodi* and *Meloidogyne hapla* on tomato plants, *Journal of Applied Phycology*, 28(3), pp. 2073–82. DOI:10.1007/s10811-015-0684-4.

Nustorova, M., Braikova, D., Gousterova, A., Vasileva-Tonkova, E. and Nedkov, P. (2006), Chemical, microbiological and plant analysis of soil fertilized with alkaline hydrolysate of sheep's wool waste, *World Journal of Microbiology and Biotechnology*, 22(4), pp. 383–90. DOI:10.1007/s11274-005-9045-9.

Oni, O. E. and Friedrich, M. W. (2017), Metal oxide reduction linked to anaerobic methane oxidation, *Trends in Microbiology*, 25(2), pp. 88–90. DOI:10.1016/j.tim.2016.12.001.

Öpik, M., Vanatoa, A., Vanatoa, E., Moora, M., Davison, J., Kalwij, J. M., Reier, Ü. and Zobel, M. (2010), The online database MaarjAM reveals global and ecosystemic distribution patterns in arbuscular mycorrhizal fungi (Glomeromycota), *New Phytologist*, 188(1), pp. 223–41. DOI:10.1111/j.1469-8137.2010.03334.x.

Ostrovskij, M. (2014), Testing HUMIN PLUS microfertilizer, *European Agrophysical Journal*, 1 (2), pp. 79–84.

Ozkan, S. and Ozkan, S. G. (2016), Investigation of humate extraction from lignites, *International Journal of Coal Preparation and Utilization*, 37(6), pp. 285–92. DOI:10.1 080/19392699.2016.1171761.

Panda, D., Pramanik, K. B. and Naya, R. (2012), Use of sea weed extracts as plant growth regulators for sustainable agriculture, *International Journal of Stress Management*, 3(3), pp. 404–11.

Pane, C., Celano, G., Villecco, D. and Zaccardelli, M. (2012), Control of *Botrytis cinerea*, *Alternaria alternata* and *Pyrenochaeta lycopersici* on tomato with whey compost-tea applications, *Crop Protection*, 38, pp. 80–6. DOI:10.1016/j.cropro.2012.03.012.

Papenfus, H. B., Kulkarni, M. G., Stirk, W. A., Finnie, J. F. and Van Staden, J. (2013), Effect of a commercial seaweed extract (Kelpak®) and polyamines on nutrient-deprived (N, P and K) okra seedlings, *Scientia Horticulturae*, 151, pp. 142–6. DOI:10.1016/j.scienta.2012.12.022.

Paredes, C., Medina, E., Bustamante, M. A. and Moral, R. (2016), Effects of spent mushroom substrates and inorganic fertilizer on the characteristics of a calcareous clayey-loam soil and lettuce production, *Soil Use and Management*, 32(4), pp. 487–94. DOI:10.1111/sum.12304.

Parvage, M., Ulén, B., Eriksson, J., Strock, J. and Kirchmann, H. (2013), Phosphorus availability in soils amended with wheat residue char, *Biology and Fertility of Soils*, 49(2), pp. 245–50. DOI:10.1007/s00374-012-0746-6.

Peterson, A. E., Walker, W. G. and Watson, K. S. (1979), Effect of whey applications on chemical properties of soils and crops, *Journal of Agricultural and Food Chemistry*, 27(4), pp. 654–8. DOI:10.1021/jf60224a064.

Qian, P., Schoenau, J. and Urton, R. (2011), Effect of soil amendment with thin stillage and glycerol on plant growth and soil properties, *Journal of Plant Nutrition*, 34(14), pp. 2206–21. DOI:10.1080/01904167.2011.618579.

Qian, S., Ding, W., Li, Y., Liu, G., Sun, J. and Ding, Q. (2015), Characterization of humic acids derived from Leonardite using a solid-state NMR spectroscopy and effects of humic acids on growth and nutrient uptake of snap bean, *Chemical Speciation & Bioavailability*, 27 (4), pp. 156–61. DOI:10.1080/09542299.2015.1118361.

Raghu, K. and MacRae, I. C. (1966), Occurrence of phosphate-dissolving micro-organisms in the rhizosphere of rice plants and in submerged soils, *Journal of Applied Bacteriology*, 29(3), pp. 582–6. DOI:10.1111/j.1365-2672.1966.tb03511.x.

Rashid, A., Mir, M. R. and Hakeem, K. R. (2016a), Biofertilizer use for sustainable agricultural production, in: Hakeem, K. R., Akhtar, M. S., and Abdullah, S. N. A. (Eds), *Plant, Soil and Microbes: Volume 1: Implications in Crop Science*. Cham, Switzerland: Springer International Publishing, pp. 163–80.

Rashid, M. I., Mujawar, L. H., Shahzad, T., Almeelbi, T., Ismail, I. M. I. and Oves, M. (2016b), Bacteria and fungi can contribute to nutrients bioavailability and aggregate formation in degraded soils, *Microbiological Research*, 183, pp. 26–41. DOI:10.1016/j.micres.2015.11.007.

Reed, E. Y., Chadwick, D. R., Hill, P. W. and Jones, D. L. (2017), Critical comparison of the impact of biochar and wood ash on soil organic matter cycling and grassland productivity, *Soil Biology and Biochemistry*, 110, pp. 134–42. DOI:10.1016/j.soilbio.2017.03.012.

Robbins, C. W. and Lehrsch, G. A. (1992), Effects of acidic cottage cheese whey on chemical and physical properties of a sodic soil, *Arid Soil Research and Rehabilitation*, 6 (2), pp. 127-34. DOI:10.1080/15324989209381305.

Roger, P. A. and Ladha, J. K. (1992), Biological N2 Fixation in wetland rice fields: Estimation and contribution to nitrogen balance, *Plant and Soil*, 141(1), pp. 41-55. DOI:10.1007/BF00011309.

Rouphael, Y., Franken, P., Schneider, C., Schwarz, D., Giovannetti, M., Agnolucci, M., Pascale, S. D., Bonini, P. and Colla, G. (2015), Arbuscular mycorrhizal fungi act as biostimulants in horticultural crops, *Biostimulants in Horticulture*, 196, pp. 91-108. DOI:10.1016/j.scienta.2015.09.002.

Roy, S., Barman, S., Chakraborty, U. and Chakraborty, B. (2015), Evaluation of spent mushroom substrate as biofertilizer for growth improvement of *Capsicum annuum* L, *Journal of Applied Biology and Biotechnology*, 3(03), pp. 22-7.

Roychowdhury, D., Paul, M. and KumarBanerjee, S. (2015), Isolation identification and characterization of phosphate solubilising bacteria from soil and the production of biofertilizer, *International Journal of Current Microbiology and Applied Sciences*, 4 (11), pp. 808-15.

Rozpara, E., Pąśko, M., Bielicki, P. and Sas Paszt, L. (2014), Influence of various bio-fertilizers on the growth and fruiting of 'Ariwa' apple trees growing in an organic orchard, *Journal of Research and Applications in Agricultural Engineering*, 59(4), pp. 65-8.

Rutkowska, B., Szulc, W., Sosulski, T. and Stepien, W. (2014), Soil micronutrient availability to crops affected by long-term inorganic and organic fertilizer applications, *Plant, Soil and Environment*, 60(5), pp. 198-203.

Saito, B. and Seckler, M. M. (2014), Alkaline extraction of humic substances from peat applied to organic-mineral fertilizer production, *Brazilian Journal of Chemical Engineering*, 31, pp. 675-82.

Sajbrt, V., Rosol, M. and Ditl, P. (2010), A comparison of distillery stillage disposal methods, *Acta Polytechnica*, 50(2), pp. 63-9.

Salama, A. S., El-Sayed, O. M. and El Gammal, O. (2014), Effect of Effective Microorganisms (EM) and potassium sulphate on productivity and fruit quality of 'Hayany' date palm grown under salinity stress, *Journal of Agriculture and Veterinary Sciences*, 7, pp. 90-9.

Salantur, A., Ozturk, A. and Akten, S. (2006), Growth and yield response of spring wheat (*Triticum aestivum* L.) to inoculation with rhizobacteria, *Plant, Soil and Environment*, 52(3), pp. 111-18.

Sanderson, K. R. and Cutcliffe, J. A. (1991), Effect of sawdust mulch on yields of select clones of lowbush blueberry, *Canadian Journal of Plant Science*, 71(4), pp. 1263-6. DOI:10.4141/cjps91-175.

Sas Paszt, L., Sumorok, B., Malusá, E., Głuszek, S. and Derkowska, E. (2011), The influence of bioproducts on root growth and mycorrhizal occurrence in the rhizosphere of strawberry plants 'Elsanta', *Journal of Fruit and Ornamental Plant Research*, 19(1), pp. 13-34.

Sas Paszt, L., Malusá, E., Sumorok, B., Canfora, L., Derkowska, E. and Głuszek, S. (2015), The influence of bioproducts on mycorrhizal occurrence and diversity in the rhizosphere of strawberry plants under controlled conditions, *Advances in Microbiology*, 5(1), pp. 40-53.

Sasirekha, B. and Srividya, S. (2016), Siderophore production by *Pseudomonas aeruginosa* FP6, a biocontrol strain for *Rhizoctonia solani* and *Colletotrichum gloeosporioides* causing diseases in chilli, *Agriculture and Natural Resources*, 50(4), pp. 250–6. DOI:10.1016/j.anres.2016.02.003.

Schäfer, T., Hanke, M.-V., Flachowsky, H., König, S., Peil, A., Kaldorf, M., Polle, A. and Buscot, F. (2012), Chitinase activities, scab resistance, mycorrhization rates and biomass of own-rooted and grafted transgenic apple, *Genetics and Molecular Biology*, 35(2), pp. 466–73. DOI:10.1590/S1415-47572012000300014.

Schlegel, A. J., Assefa, Y., Bond, H. D., Haag, L. A. and Stone, L. R. (2017), Changes in soil nutrients after 10 years of cattle manure and swine effluent application, *Soil and Tillage Research*, 172, pp. 48–58. DOI:10.1016/j.still.2017.05.004.

Shaheen, S., Khan, M., Khan, M. J., Jilani, S., Bibi, Z., Munir, M. and Kiran, M. (2017), Effective Microorganisms (EM) co-applied with organic wastes and NPK stimulate the growth, yield and quality of spinach (*Spinacia oleracea* L.), *Sarhad Journal of Agriculture*, 33 (1), pp. 30–41.

Sharp, G. R. (2013), A review of the applications of chitin and its derivatives in agriculture to modify plant- microbial interactions and improve crop yields, *Agronomy*, 3(4), pp. 778–93. DOI:10.3390/agronomy3040757.

Sharratt, W. J., Peterson, A. E. and Calbert, H. E. (1959), Whey as a source of plant nutrients and its effect on the soil, *Journal of Dairy Science*, 42(7), pp. 1126–31. DOI:10.3168/jds.S0022-0302(59)90705-2.

Siebers, N., Godlinski, F. and Leinweber, P. (2014), Bone char as phosphorus fertilizer involved in cadmium immobilization in lettuce, wheat, and potato cropping, *Journal of Plant Nutrition and Soil Science*, 177(1), pp. 75–83. DOI:10.1002/jpln.201300113.

Sim, E. Y. S. and Wu, T. Y. (2010), The potential reuse of biodegradable municipal solid wastes (MSW) as feedstocks in vermicomposting, *Journal of the Science of Food and Agriculture*, 90(13), pp. 2153–62. DOI:10.1002/jsfa.4127.

Singh, R., Gupta, R. K., Patil, R. T., Sharma, R. R., Asrey, R., Kumar, A. and Jangra, K. K. (2010), Sequential foliar application of vermicompost leachates improves marketable fruit yield and quality of strawberry (*Fragaria* × *ananassa* Duch.), *Scientia Horticulturae*, 124(1), pp. 34–9. DOI:10.1016/j.scienta.2009.12.002.

Singh, P., Singh, M. and Tripathi, B. (2013), Glomalin: An arbuscular mycorrhizal fungal soil protein, *Protoplasma*, 250(3), pp. 663–9. DOI:10.1007/s00709-012-0453-z.

Singh, A. K., Beer K., and Kumar Pal, A. K. (2015), Effect of vermicompost and biofertilizers on strawberry I: Growth, flowering and yield, *Annals of Plant and Soil Research*, 17, pp. 196–9.

Singh, A., Kaushik, M. S., Srivastava, M., Tiwari, D. N. and Mishra, A. K. (2016), Siderophore mediated attenuation of cadmium toxicity by paddy field cyanobacterium *Anabaena oryzae*, *Algal Research*, 16, pp. 63–8. DOI:10.1016/j.algal.2016.02.030.

Sohi, S. P., Krull, E., Lopez-Capel, E. and Bol, R. (2010), A review of biochar and its use and function in soil, *Advances in Agronomy*, 105, pp. 47–82. DOI:10.1016/S0065-2113(10)05002-9.

Souza, E. M., Chubatsu, L. S., Huergo, L. F., Monteiro, R., Camilios-Neto, D., Wassem, R. and de Oliveira Pedrosa, F. (2014), Use of nitrogen-fixing bacteria to improve agricultural productivity, *BMC Proceedings*, 8(4), pp. O23. DOI:10.1186/1753-6561-8-S4-O23.

Spokas, K. A. (2010), Review of the stability of biochar in soils: Predictability of O:C molar ratios, *Carbon Management*, 1(2), pp. 289–303. DOI:10.4155/cmt.10.32.

Spokas, K. A., Koskinen, W. C., Baker, J. M. and Reicosky, D. C. (2009), Impacts of woodchip biochar additions on greenhouse gas production and sorption/degradation of two herbicides in a Minnesota soil, *Chemosphere*, 77(4), pp. 574–81. DOI:10.1016/j. chemosphere.2009.06.053.

Spokas, K. A., Cantrell, K. B., Novak, J. M., Archer, D. W., Ippolito, J. A., Collins, H. P., Boateng, A. A., Lima, I. M., Lamb, M. C., McAloon, A. J., Lentz, R. D. and Nichols, K. A. (2012a), Biochar: A synthesis of its agronomic impact beyond carbon sequestration, *Journal of Environmental Quality*, 41(4), pp. 973–89.

Spokas, K., Novak, J. and Venterea, R. (2012b), Biochar's role as an alternative N-fertilizer: Ammonia capture, *Plant and Soil*, 350(1–2), pp. 35–42. DOI:10.1007/s11104-011-0930-8.

Stadnik, M. J. and Freitas, M. B. de (2014), Algal polysaccharides as source of plant resistance inducers, *Tropical Plant Pathology*, 39(2), pp. 111–18.

Stefaniuk, M., Bartmiński, P., Różyło, K., Dębicki, R. and Oleszczuk, P. (2015), Ecotoxicological assessment of residues from different biogas production plants used as fertilizer for soil, *Journal of Hazardous Materials*, 298, pp. 195–202. DOI:10.1016/j. jhazmat.2015.05.026.

Stefaniuk, M., Oleszczuk, P. and Bartmiński, P. (2016), Chemical and ecotoxicological evaluation of biochar produced from residues of biogas production, *Journal of Hazardous Materials*, 318, pp. 417–24. DOI:10.1016/j.jhazmat.2016.06.013.

Steiner, C., Das, K. C., Garcia, M., Förster, B. and Zech, W. (2008), Charcoal and smoke extract stimulate the soil microbial community in a highly weathered xanthic Ferralsol, *Pedobiologia*, 51(5–6), pp. 359–66. DOI:10.1016/j.pedobi.2007.08.002.

Steiner, C., Garcia, M. and Zech, W. (2009), Effects of charcoal as slow release nutrient carrier on N-P-K dynamics and soil microbial population: Pot experiments with Ferralsol substrate, in: Woods, W., Teixeira, W., Lehmann, J., Steiner, C., WinklerPrins, A., and Rebellato, L. (Eds), *Amazonian Dark Earths: Wim Sombroek's Vision*. Dordrecht, the Netherlands: Springer Netherlands, pp. 325–38.

Stępień, W., Malusà, E., Paszt, L. S., Renzi, G. and Ciesielska, J. (2012), Effect of brown coal- based composts produced with the use of white rot fungi on the growth and yield of strawberry plants, in: Ecofruit. *15th International Conference on Organic Fruit-Growing. Proceedings for the Conference*. Hohenheim, Germany: Fördergemeinschaft Ökologischer Obstbau eV (FÖKO), pp. 302–6.

Stirk, W. A. and Staden, J. (1996), Comparison of cytokinin- and auxin-like activity in some commercially used seaweed extracts, *Journal of Applied Phycology*, 8(6), pp. 503–8. DOI:10.1007/BF02186328.

Struszczyk, H., Pospieszny H., and Kotliński S. (1989), Some new application of chitosan, in: Skjoak-Bræk G., Anthonsen T., Sandford P. (Eds), *Chitin and Chitosan*, Elsevier Science, London. pp. 733–42.

Sumbul, A., Rizvi, R., Mahmood, I. and Ansari, R. A. (2015), Oil-cake amendments: Useful tools for the management of phytonematodes, *Asian Journal of Plant Pathology*, 9(3), pp. 91–111.

Sun, Q., Ding, W., Yang, Y., Sun, J. and Ding, Q. (2016), Humic acids derived from leonardite-affected growth and nutrient uptake of corn seedlings, *Communications in Soil Science and Plant Analysis*, 47(10), pp. 1275–82. DOI:10.1080/00103624.2016.1178767.

Szajdak, L. W. (2016), Phytohormone in peats, sapropels, and peat substrates, in: Szajdak, L.W. (Ed.), *Bioactive Compounds in Agricultural Soils*. Cham: Springer International Publishing, pp. 247–72.

Szpak, P., Longstaffe, F. J., Millaire, J.-F. and White, C. D. (2012), Stable isotope biogeochemistry of seabird guano fertilization: Results from growth chamber studies with maize (*Zea mays*), *PLoS ONE*, 7 (3), pp. e33741 (1–16). DOI:10.1371/journal.pone.0033741.

Taktek, S., Trépanier, M., Servin, P. M., St-Arnaud, M., Piché, Y., Fortin, J.-A. and Antoun, H. (2015), Trapping of phosphate solubilizing bacteria on hyphae of the arbuscular mycorrhizal fungus Rhizophagus irregularis DAOM 197198, *Soil Biology and Biochemistry*, 90, pp. 1–9. DOI:10.1016/j.soilbio.2015.07.016.

Tejeda-Agredano, M.-C., Mayer, P. and Ortega-Calvo, J.-J. (2014), The effect of humic acids on biodegradation of polycyclic aromatic hydrocarbons depends on the exposure regime, *Environmental Pollution*, 184, pp. 435–42. DOI:10.1016/j.envpol.2013.09.031.

Thorsen, M., Woodward, S. and McKenzie, B. (2010), Kelp (*Laminaria digitata*) increases germination and affects rooting and plant vigour in crops and native plants from an arable grassland in the Outer Hebrides, Scotland, *Journal of Coastal Conservation*, 14(3), pp. 239–47. DOI:10.1007/s11852-010-0091-6.

Toselli, M., Sorrenti, G., Marangoni, B., Innocenti, A., Baldi, E., Marcolini, G. and Quartieri, M. (2013), Effect of organic fertilization on soil fertility, tree nutritional status and nutrient removal of mature nectarine trees, *Acta Horticulturae*, 1001, pp. 303–10.

Tretjakova, R., Grebeža, J. and Martinovs, A. (2015), Research into biological characteristics of dried sapropel, in: *Environment. Technology, Resources*. Rezekne, Latvia, 1, pp. 223–7.

Troy, S. M., Nolan, T., Kwapinski, W., Leahy, J. J., Healy, M. G. and Lawlor, P. G. (2012), Effect of sawdust addition on composting of separated raw and anaerobically digested pig manure, *Journal of Environmental Management*, 111, pp. 70–7. DOI:10.1016/j.jenvman.2012.06.035.

Vaneeckhaute, C., Meers, E., Michels, E., Buysse, J. and Tack, F. M. G. (2013), Ecological and economic benefits of the application of bio-based mineral fertilizers in modern agriculture, *Biomass and Bioenergy*, 49, pp. 239–48. DOI:10.1016/j.biombioe.2012.12.036.

Vanek, S. J., Thies, J., Wang, B., Hanley, K. and Lehmann, J. (2016), Pore-size and water activity effects on survival of Rhizobium tropici in biochar inoculant carriers, *Journal of Microbial & Biochemical Technology*, 8, pp. 296–306.

Vasconcellos, R. L. F., Bonfim, J. A., Baretta, D. and Cardoso, E. J. B. N. (2016), Arbuscular mycorrhizal fungi and glomalin-related soil protein as potential indicators of soil quality in a recuperation gradient of the atlantic forest in Brazil, *Land Degradation & Development*, 27(2), pp. 325–34. DOI:10.1002/ldr.2228.

Velázquez, E., Carro, L., Flores-Félix, J. D., Martínez-Hidalgo, P., Menéndez, E., Ramírez-Bahena, M.-H., Mulas, R., González-Andrés, F., Martínez-Molina, E. and Peix, A. (2017), The legume nodule microbiome: A source of plant growth- promoting bacteria, in: Kumar, V., Kumar, M., Sharma, S., and Prasad, R. (Eds), *Probiotics and Plant Health*. Singapore: Springer Singapore, pp. 41–70.

Vessey, J. K. (2003), Plant growth promoting rhizobacteria as biofertilizers, *Plant and Soil*, 255(2), pp. 571–86. DOI:10.1023/A:1026037216893.

Vicario, J. C., Primo, E. D., Dardanelli, M. S. and Giordano, W. (2016), Promotion of peanut growth by co- inoculation with selected strains of Bradyrhizobium and Azospirillum, *Journal of Plant Growth Regulation*, 35(2), pp. 413–19. DOI:10.1007/s00344-015-9547-0.

Walpola, B. C. and Yoon, M.-H. (2012), Prospectus of phosphate solubilizing microorganisms and phosphorus availability in agricultural soils: A review, *African Journal of Microbiology Research*, 6(37), pp. 6600–5. DOI:10.5897/AJMR12.889

Wani, S. P. and Lee, K. K. (2002), Population dynamics of nitrogen fixing bacteria associated with pearl millet (*P. americanum* L.), in: *Biotechnology of Nitrogen Fixation in the Tropics*. University of Pertanian, Malaysia, pp. 21–30.

Wani, F. S., Latief Ahmad, T. A. and Mushtaq, A. (2015), Role of microorganisms in nutrient mobilization and soil health-A review, *Journal of Pure and Applied Microbiology*, 9(2), pp. 1401–10.

Wilkie, A. C., Riedesel, K. J. and Owens, J. M. (2000), Stillage characterization and anaerobic treatment of ethanol stillage from conventional and cellulosic feedstocks, *Biomass and Bioenergy*, 19(2), pp. 63–102. DOI:10.1016/S0961-9534(00)00017-9.

Yang, X., Wang, X., Wang, K., Su, L., Li, H., Li, R. and Shen, Q. (2015a), The nematicidal effect of Camellia seed cake on root-knot nematode Meloidogyne javanica of banana, *PLoS ONE*, 10 (4), pp. e0119700 (1–18). DOI:10.1371/journal.pone.0119700.

Yang, L., Zhao, F., Chang, Q., Li, T. and Li, F. (2015b), Effects of vermicomposts on tomato yield and quality and soil fertility in greenhouse under different soil water regimes, *Agricultural Water Management*, 160, pp. 98–105. DOI:10.1016/j.agwat.2015.07.002.

Yooyongwech, S., Samphumphuang, T., Tisarum, R., Theerawitaya, C. and Cha-um, S. (2016), Arbuscular mycorrhizal fungi (AMF) improved water deficit tolerance in two different sweet potato genotypes involves osmotic adjustments via soluble sugar and free proline, *Scientia Horticulturae*, 198, pp. 107–17. DOI:10.1016/j.scienta.2015.11.002.

Yu, T., Wang, L., Yin, Y., Wang, Y. and Zheng, X. (2008), Effect of chitin on the antagonistic activity of *Cryptococcus laurentii* against *Penicillium expansum* in pear fruit, *International Journal of Food Microbiology*, 122(1-2), pp. 44–8. DOI:10.1016/j.ijfoodmicro.2007.11.059

Yu, G., Ran, W. and Shen, Q. (2016), Compost process and organic fertilizers application in China, in: Larramendy, M. L. and Soloneski, S. (Eds), *Organic Fertilizers – From Basic Concepts to Applied Outcomes*. Rijeka: InTech, pp. 1–24.

Yusuf, R., Kristiansen, P. and Warwick, N. (2016), Effect of two seaweed products on radish (*Raphanus sativus*) growth under greenhouse conditions, *AGROLAND: The Agricultural Sciences Journal*, 2(1), pp. 1–7.

Zeng, D., Luo, X. and Tu, R. (2012), Application of bioactive coatings based on chitosan for soybean seed protection, *International Journal of Carbohydrate Chemistry*, 2012, pp. 104565 (1–5).

Zhang, H., Ding, W., He, X., Yu, H., Fan, J. and Liu, D. (2014), Influence of 20-year organic and inorganic fertilization on organic carbon accumulation and microbial community structure of aggregates in an intensively cultivated sandy loam soil, *PLoS ONE*, 9(3), pp. e92733. DOI:10.1371/journal.pone.0092733.

Zhao, R., Guo, W., Bi, N., Guo, J., Wang, L., Zhao, J. and Zhang, J. (2015), Arbuscular mycorrhizal fungi affect the growth, nutrient uptake and water status of maize (*Zea mays* L.) grown in two types of coal mine spoils under drought stress, *Applied Soil Ecology*, 88, pp. 41-9. DOI:10.1016/j.apsoil.2014.11.016.

Zhen, Z., Liu, H., Wang, N., Guo, L., Meng, J., Ding, N., Wu, G. and Jiang, G. (2014), Effects of manure compost application on soil microbial community diversity and soil microenvironments in a temperate cropland in China, *PLoS ONE*, 9(10), pp. e108555. DOI:10.1371/journal.pone.0108555.

Chapter 2

Optimizing the use of treated wastes in crop nutrition

Sylvia Kratz, Kerstin Panten, Ewald Schnug and Elke Bloem, Julius Kühn-Institute, Germany

1 Introduction

A rapid population growth and urbanization over the last century has brought with it an enormous increase in the production of human wastes. At the same time, human activities have caused considerable disturbance and unbalancing of natural global nutrient flows. In response to this, we nowadays observe intensified efforts to close nutrient cycles, which are mirrored in a growing number of supranational conventions and legal frameworks aimed at a sustainable management and reuse or recycling of human wastes (e.g. the EU Circular Economy Action Plan, European Commission, 2014a, 2015 and its related yearly Circular Economy Packages).

The most important reasons for this are environmental concerns because nutrient losses from waste materials cause eutrophication and pollution of ground and surface water bodies and the environment. Another important point is the fact that especially for phosphorus (P) the global reserves are becoming increasingly depleted, and there is no replacement for P in crop nutrition, endangering food production for the growing population. While nitrogen (N) for fertilizer production can be recovered synthetically from the atmosphere and is therefore a mostly unlimited resource, global reserves of phosphate rock (the raw material used to produce phosphate fertilizers) are non-renewable. Their depletion has been subject to intensive discussions during the last decade ('Peak Phosphorus'; Cordell and White, 2011). Furthermore, 85% of the

http://dx.doi.org/10.19103/AS.2019.0062.32

known P reserves are situated in only three countries, namely Morocco, China and the United States (Lou et al., 2018), with Morocco being the main global exporter (Schoumans et al., 2015). In contrast to this, the European Union (EU) only has very limited phosphate rock reserves at its disposal, which by no means meet the internal demands for P (Vollaro et al., 2016). This has even led to the inclusion of mineral phosphate rock and P on the list of critical raw materials of the European Union (EU) (European Commission, 2014b, 2017). On the other hand, many human wastes contain large amounts of essential plant nutrients, making them interesting sources for crop nutrition in agriculture. The potential of human waste streams to be turned into fertilizers is often quantified in the first place with regard to their P replacement potential (e.g. van Dijk et al., 2016; Möller et al., 2018). Also, development of techniques for treatment or recovery of nutrients from waste materials has mainly focussed on P. In line with this, the selection of treated wastes and presentation of relevant recycling and recovery techniques in this chapter will also focus on P. Wastes containing secondary macronutrients only (e.g. sulphur-containing flue gas ashes or products from desulphurization of natural gas, diesel fuel and fuel oils) will not be discussed here.

Several types of treated wastes contain substantial amounts of P and are therefore regarded as potentially relevant to crop nutrition. These include residues from waste water treatment (sewage sludge), slaughtering wastes as well as 'biowastes' such as biodegradable garden and park waste, food and kitchen waste from households, restaurants, caterers and retail premises and comparable waste from food-processing plants. Biowastes are usually composted or digested before usage in fertilization (van Dijk et al., 2016; Möller et al., 2018; Weissengruber et al., 2018). In contrast to that, animal manures and products derived from them are generally not considered 'wastes', but – despite the difficulties related to their equitable distribution and sustainable use – are perceived as valuable organic fertilizers, which offer a direct opportunity to close nutrient cycles on farm. Their optimum use in crop nutrition is the topic of a separate chapter in this book.

In order to serve as an efficient and environmentally sustainable fertilizer, a waste material must not only have a relevant nutrient content, but nutrients must be provided in a plant-available form, and the material must be free of substances that may have adverse effects on the health of plants and soils, animals, humans and the surrounding environment. The aim of this chapter is to give an overview of aspects that need to be taken into consideration regarding the use of treated wastes to replace mineral P fertilizers. Sewage sludge disposal to agriculture is a matter of highly controversial debate today, mainly with respect to its high risk of organic and inorganic contamination. Different strategies are under evaluation to recover P from this waste stream in the plant-available forms. Therefore, in this chapter the optimized use of treated wastes is discussed by taking sewage sludge as an example. However, many of the key issues highlighted here would be applicable also to other human waste streams with fertilizing potential.

2 Key issues for the optimum use of treated wastes in crop nutrition

Different aspects are of importance when evaluating treated wastes with respect to their fertilizing potential in agriculture (Fig. 1). First of all, nutrient content and plant availability are of relevance for a growing crop; in addition, organic matter can be important for improving soil organic matter (SOM) content and related soil functions, such as the capacity to store and filter water, the production of food and fibre as a source of nutrients and habitat for soil organisms or to sequester carbon. Moreover, transportation costs are of relevance because only materials with high nutrient and dry-matter (DM) content are worthy of being transported over longer distances. Also of superior importance is the possible contamination by harmful substances and organisms that need to be evaluated before treated waste can be recommended for agricultural use. Treated wastes can be like a black box containing an unknown amount of contaminants. Some of these aspects will be addressed in the following sections using sewage sludge as an example. In Section 3, the key factors for direct land application of sewage sludge will be discussed. Section 4 looks at further processing and products derived from waste water and sewage sludge.

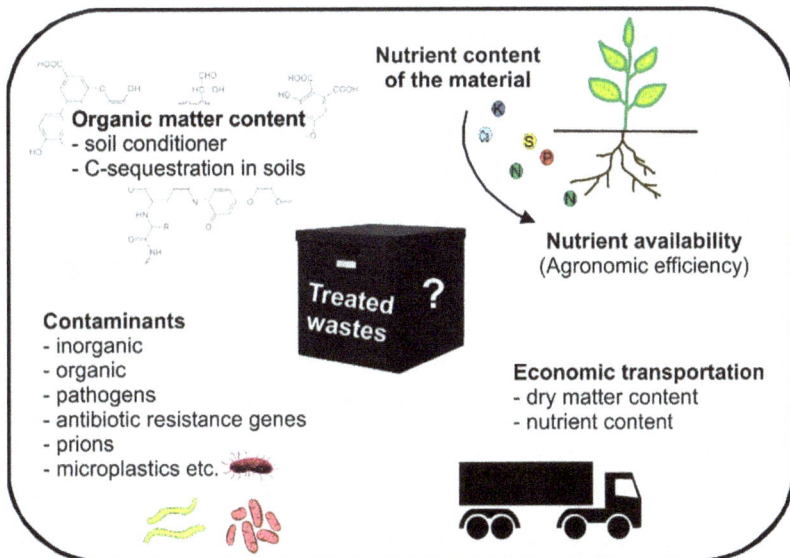

Figure 1 Key issues in the evaluation of use of treated wastes in plant nutrition.

3 Direct land application of sewage sludge

The aim of municipal waste water treatment is to produce a clean effluent that may be discharged back into the environment. To this end, incoming waste water is cleaned in the waste water treatment plants (WWTP) in three treatment steps. The primary (mechanical) treatment step involves screening for coarse solids and grit removal (sand-like solids), followed by gravity sedimentation to remove suspended solids. In the secondary (biological) treatment step, microorganisms are used to reduce biochemical oxygen demand. The tertiary treatment step includes biological and/or chemical precipitation processes to remove N and P from the waste water (Epstein, 2003). Over the last few years, a fourth treatment step has been investigated and implemented by a number of modern WWTPs, aiming to free the cleaned water from unwanted organic micropollutants such as pharmaceuticals, personal-care products and hormones. Different techniques, including membrane separation, activated carbon adsorption, advanced oxidation processes/ozonation and UV photolysis, have been tested (Ahmed et al., 2017; Grandclément et al., 2017; Alvarino et al., 2018).

The residue left after these treatment steps is known as *sewage sludge*. Before it can be used as a fertilizer, raw/activated sewage sludge is usually dewatered (and/or dried) and stabilized in some way in order to reduce bulk and generate a stable material that does not pose a risk to human health or the environment. To this end, biological techniques such as anaerobic/aerobic digestion or composting, chemical processes (alkaline stabilization/lime treatment) and/or physical processes such as pasteurization, thermal hydrolysis, thermal drying or air/solar drying are applied (Epstein, 2003; Rigby et al., 2016). Treated sewage sludge is often termed 'biosolids' in the international literature, which is in line with the definitions laid down in the US legislation. Following Verlicchi and Zambello (2015), we use the term 'sewage sludge' in a broad sense here to denote a mixture of the residuals from WWTPs, which can be primary, secondary or treated sludge.

Direct land application of sewage sludge as a fertilizer is still a common practice in many parts of the world. It is associated with a number of advantages, most prominently the supply of essential plant nutrients such as N and P, and the addition of organic matter to the soil, which may improve the soil structure and could increase the availability of nutrients (Epstein, 2003; Goldbach et al., 2007; Singh and Agrawal, 2008; Hemmat et al., 2010; Kominko et al., 2017; Börjesson and Kätterer, 2018). More than 30 years ago, the *EU Council Directive 86/278/EEC* of 12 June 1986 on the protection of the environment, and in particular of the soil, when sewage sludge is used in agriculture (which was considered for revision in 2010, however, without a final result, see European Commission, 2016) was designed to encourage the controlled and efficient reuse of sewage sludge by land application in agriculture, while regulating it in a way that should prevent adverse effects on soil, water, vegetation, humans and animals. Similarly, in the United States, after 10 years of intensive research and risk assessment, standards for the use or disposal of sewage sludge were implemented as *Part 503 of the USEPA Biosolids Rule* in 1993 (amended in 1994) to promote the beneficial and environmentally safe use of sewage sludge in agriculture. The USEPA Rule *503* legally defined the term 'biosolids' for sludge that has undergone stabilizing treatment such as anaerobic or aerobic digestion, alkaline stabilization, composting or heat drying and pelletizing and can therefore be considered fit for land application, given it is under the limit values for inorganic and organic contaminants set by this rule (Epstein, 2003). Comparable legislation or guidelines were introduced in other parts of the world such as Canada (Webber and Sidhwa, 2007; CCME, 2010), Australia and New Zealand (NWQMS, 2004).

Around 30 years later, the perception of land application of sewage sludge has changed considerably in some parts of the world. In Europe, for example, around 45% of the total sewage sludge produced was used in agriculture in 2012 (Fig. 2). However, while some countries such as Denmark, France, Spain, Portugal, the United Kingdom, Ireland, Italy and Norway still rely strongly on agricultural use as the preferred disposal route for sewage sludge, others such as the Netherlands, Belgium, Germany and Switzerland have either completely banned this practice already or are at least 'en route' towards such a ban in

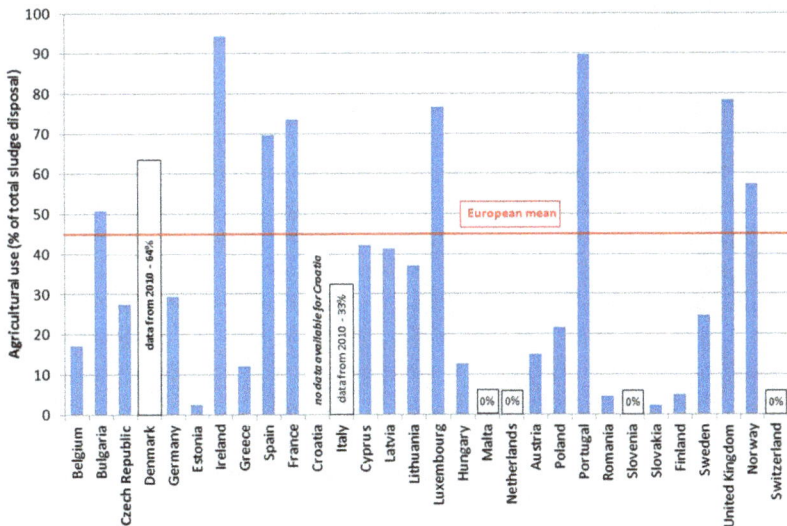

Figure 2 Agricultural use of sewage sludge in Europe (EU-28, Norway and Switzerland) in 2012. Source: data from EUROSTAT (2018), note: data sets for years 2013–15 are too incomplete for calculations; no data available for Croatia for 2012.

the near future, and are turning towards other disposal routes such as sludge incineration as the prevalent practice (Mininni et al., 2015).

In other high-income countries, such as the United States, Canada, Australia and New Zealand, direct land application of sewage sludge is still widely practiced; however, these countries are also looking at sludge incineration as a viable and promising alternative disposal route (Rigby et al., 2016; Fijalkowski et al., 2017). This is different in many low- or middle-income countries, who perceive sewage sludge as an indispensable and valuable source of nutrients and organic matter (e.g. Singh and Agrawal, 2008; Melo et al., 2018; Hamdi et al., 2019).

There are a number of reasons that have been brought forward against direct land application; the most important ones will be discussed in the following sections.

3.1 Limited plant availability of nutrients

Kirchmann et al. (2017) and Börjesson and Kätterer (2018) argued that the plant availability of N and P from sewage sludge was rather limited.

Median *total N contents* in different types of sewage sludge reported in an extensive international literature review (Rigby et al., 2016) ranged between 15 and 61 g/kg DM. Of the total N content, between 3% and up to 43% is present in plant-available inorganic forms, that is mainly as ammonium N, whereas the rest is organically bound and needs to be mineralized before it can be taken up by plants. The large variation in the size of the inorganic N fraction is mainly due to sludge treatment (Table 1). Similarly to the proportion of inorganic N, the rate of mineralization of the organic N compounds also depends on the type of sludge treatment (dewatering/drying and stabilization), as well as on soil parameters such as soil type, moisture and pH and on climatic conditions and their influence on soil microbial activity. Thus, the fraction of mineralizable N for sewage sludge may range between 7% of total organic N for composted sewage sludge and up to 47% of total organic N for microbiologically treated, aerobically digested sewage sludge (Rigby et al., 2016, see Table 1). Evidence from the literature reviewed by Rigby et al. (2016) suggested that – in contrast to earlier research – at regular agronomic application rates based on crop requirements and regulatory restrictions, most of the mineralizable organic N fraction will be rapidly released and available for crop uptake within the first year after application. A residual N value may only occur due to unfavourable environmental factors that slow down or prevent microbial mineralization, such as dry soil conditions and/or cool winter temperatures.

Based on long-term field trials, some researchers claim that, once an equilibrium between addition and mineralization of organic substance is

Table 1 Typical median values of total and inorganic N contents and the mean and the range of mineralizable N fraction in different types of sewage sludge

Type of sludge treatment	Total N (g/kg DM)	Inorganic N (% of total N)	Mineralizable N (% of organic N)
Activated (raw/unstabilized)			
Liquid	48	15.8	No data
Dewatered	61	3.2	42 (20-64)
Air dried	47	4.9	14 (0-30)
After aerobic digestion			47 (32-58)
Liquid	58	4.7	
Dewatered	53	4.4	
Air dried	28	4.4	
After mesophilic anaerobic digestion			30 (11-45)
Liquid	53	42.7	
Dewatered	53	14.9	
Air dried	23	12.8	
Dewatered (after thermophilic aerobic-anaerobic digestion)	47	28.7	34 (33-37)
Thermally dried	54	4.3	40 (26-71)
Lime treated (alkaline stabilization)			34 (2.6-65)
Liquid	35	4.0	
Dewatered	37	4.2	
Air dried	15	4.9	
Composted	16	5.2	6.7 (10-24)

Source: modified after Rigby et al. (2016).

reached (which may take around 40-50 years), N efficiency/fertilizer replacement potential of sewage sludge may be roughly similar to that of mineral N fertilizers (Goldbach et al., 2007). In contrast, others observed a low N-use efficiency of sewage sludge (between 3% and 21%) in their field trials, even after 20 or more years (Kirchmann et al., 2017; Börjesson and Kätterer, 2018). Such contradictory results may well be the consequence of the large variations in N content and quality occurring among different types of sludge, as well as of differences in experimental conditions such as soil types and climatic influences. Therefore, Rigby et al. (2016) emphasized that, as a basis for the implementation of best management practices for land application of sewage sludge, estimates of N fertilizer replacement values should account for local/regional variations in waste water and sludge treatment processes, as well as for the effects of soil environmental conditions that may influence the size of the mineralizable N pool and the extent of its mineralization.

Total P contents in sewage sludge may also vary, depending on the type of treatment as well as on the origin of the incoming waste water, typically ranging around 10–40 mg/kg DM (Table 2). The plant availability of the total P content depends on the P forms found in the sludge, which are determined by the type of P removal practiced in the WWTPs. Between 60% and 90% of the P occur in inorganic forms (Kratz et al., 2016). The share of inorganic forms is higher when chemical P precipitation is used for P removal (Steckenmesser et al., 2017). Most commonly, inorganic Fe and Al salts in chloride or sulphate form are used to precipitate P from the waste water, resulting in the majority of P being strongly bound as Fe/Al phosphates, or chemisorbed to Fe and/or Al in various types of (amorphous) Fe/Al-(hydroxy)oxides. In addition, stable Ca phosphates such as the sparingly or insoluble hydroxyapatite and tricalcium phosphate (whitlockite), as well as some dicalcium phosphate, the latter being known to be plant available, are found (Frossard et al., 1994, 1996, 1997; Huang and Shenker, 2004; Shober et al., 2006; Steckenmesser et al., 2017). If CaO is used for sludge stabilization, non-crystalline nanoparticles of Ca phosphate phases, coated with adsorbed organic matter, may also occur (Huang and Shenker, 2004).

Table 2 Share of easily exchangeable P fraction (EEF) (% of total P) for different types of sewage sludge

Type of P removal	Total P (g/ kg DM)	P-fraction (method)	EEF (% of total P)
No chemical precipitation[b]	15	NH_4Cl-P (Psenner et al., 1984)	13.4
No chemical precipitation[b]	10	NH_4Cl-P (Psenner et al., 1984)	2.6
EBPR[*] + $Al_2(SO_4)_3$[d]	13	NH_4Cl-P (Chang and Jackson, 1957)	3.3
BPR[*] + lime[a]	20	NH_4Cl-P (Kuo, 1996)	8.9
$FeSO_4$[b]	12	NH_4Cl-P (Psenner et al., 1984)	15.8
$CaCl_2$[b]	17	NH_4Cl-P (Psenner et al., 1984)	5.2
$FeCl_2$[d]	18	NH_4Cl-P (Chang and Jackson, 1957)	2.5
Polyaluminum chloride and/or Fe-chloride and/or Fe/Al-sulfate[c]	11–37	$NaHCO_3$-P (Hedley et al., 1982)	<5
Polyaluminum chloride + Fe-chloride + lime[c]	7	$NaHCO_3$-P (Hedley et al., 1982)	38
Polyaluminum chloride + Fe-chloride + lime[c]	16	$NaHCO_3$-P (Hedley et al., 1982)	17
Lime + Al[a]	15	NH_4Cl-P (Kuo, 1996)	4.2
Lime + Fe[a]	13	NH_4Cl-P (Kuo, 1996)	3.0
$FeCl_3$[a]	20	NH_4Cl-P (Kuo, 1996)	1.1
$FeCl_3$[a]	27	NH_4Cl-P (Kuo, 1996)	0.3

[*] (E)BPR = (enhanced) biological P removal.
References: [a] Shober et al. (2006), [b] Xu et al. (2012), [c] Øgaard and Brod (2016), [d] Steckenmesser et al. (2017).

Organic P forms found in sewage sludge are the monoester inositol phosphate/phytic acid, diesters (phospholipids), long-chain polyphosphates and pyrophosphates (Huang and Tang, 2015; Kratz et al., 2016). If enhanced biological P removal (EBPR) is applied in the WWTPs, the share of organic polyphosphates increases considerably (Steckenmesser et al., 2017). In this case, phosphate is released from the activated sludge in an anaerobic zone followed by a 'luxury uptake' of phosphate by microorganisms in the aerated and anoxic zones of this special reactor configuration. Phosphorus taken up by those microorganisms is mainly internally stored in the cells in the form of long-chain polyphosphate granules. Long-chain polyphosphates are labile and may hydrolyse easily under acidic or alkaline conditions, forming short-chain polyphosphates, pyrophosphates or orthophosphates (Zhang et al., 2013; Huang and Tang, 2015; Kratz et al., 2019).

The origin of the waste water (e.g. share of domestic vs. industrial waste water) also plays an important role in P speciation because it determines the chemical composition of the sludge, including the concentrations of Fe and Al, and thus the molar ratios Fe/P and Al/P (Xu et al., 2012). Several studies have indicated that the molar ratio (Fe + Al)/P is the key factor determining P speciation and thus solubility/plant availability in sewage sludge (Römer and Samie, 2001; Miller and O'Connor, 2009; Xu et al., 2012; Kahiluoto et al., 2015; Øgaard and Brod, 2016). Øgaard and Brod (2016) demonstrated that precipitation with Al salts had a stronger effect in decreasing P plant availability than precipitation with Fe salts, which was explained by the lower solubility of Al phosphates in comparison to Fe-phosphates. According to Kahiluoto et al. (2015), who studied P efficiency of sludges with three different Fe/P ratios (biologically treated sludge: Fe/P < 0.2; Fe-sludge #1: Fe/P 1.6; Fe-sludge #2: Fe/P 9.8), sludge P might even reach plant availability of the same order of magnitude as a soluble mineral NPK fertilizer, if the Fe/P ratio were 1.6 or lower. This is in line with O'Connor et al. (2004) and Miller and O'Connor (2009), who found P uptake of the same order of magnitude as from water-soluble triple superphosphate or even higher for sewage sludges from biological P removal (BPR) with a low molar ratio of oxalate-soluble (Fe + Al)/P (between 0.3 and 0.8), whereas sludges with molar ratios above 1 performed significantly worse. On the other hand, Øgaard and Brod (2016) reported rather low mineral fertilizer equivalents (MFE) (based on P uptake: 5–24% for a short term (first cut), increasing up to 19–52% for the sum of cuts 3–6 in their 26-week pot trial with ryegrass) for chemically precipitated sludges with molar ratios (Fe + Al)/P between 2.8 and 13. As pointed out by Römer and Samie (2001) and confirmed by Huang et al. (2012), a higher molar ratio of Fe/P would have not only a negative influence on P speciation in the sludge itself, but might also lead to an increased P sorption capacity of agricultural soils due to the excess

Fe applied to soil, thereby decreasing the concentration of plant-available orthophosphate in the soil solution.

Due to the many different inorganic P species of different solubility that may occur in sewage sludge, assessment of the share of plant-available P by chemical extraction methods is not as straightforward as for N. A number of researchers have carried out sequential fractionation studies (e.g. Shober et al., 2006; Xu et al., 2012; Øgaard and Brod, 2016; Steckenmesser et al., 2017) to determine the share of easily exchangeable ('labile') P forms in sewage sludge, which are thought to be readily plant available. Mainly, NH_4Cl or $NaHCO_3$ are used to extract this fraction. The results may range between less than 1% and up to 38% of total P (Table 2). Despite the fact that EBPR sludge has a higher share of easily hydrolysable organic polyphosphates, the 'labile' P fraction as determined by sequential fractionation is not necessarily higher than for chemically precipitated sludge. This may be due to the fact that EBPR facilities are usually equipped with an additional chemical precipitation in order to guarantee the legally defined target effluent values for phosphate (Shober et al., 2006; Kratz et al., 2019) or, as indicated before, due to a high molar ratio (Fe + Al)/P of the incoming waste water. The lime treatment may increase considerably the labile P fraction in chemically precipitated sludge, as well as plant P uptake from such sludge (Øgaard and Brod, 2016); however, the effect of liming on plant P uptake appears to depend on the molar ratio Ca/P, which has an impact on the type and stability of the Ca phosphates formed in the sludge (Kahiluoto et al., 2015).

While acknowledging the fertilizing potential of sewage sludge in principle, some authors (e.g. Huang and Shenker, 2004; Xu et al., 2012; Kominko et al., 2017) mention its low *N/P ratio* as a limitation for its direct use as a fertilizer because sludge applications based on the N requirements of crops may lead to over-application of P, resulting in P losses and water eutrophication. This is confirmed by the long-term data on N and P contents in German sewage sludge applied in agriculture between 2001 and 2015: total N contents ranged between 39 and 45 g/kg DM and total P contents between 22 and 27 g/kg DM, resulting in the range of N/P ratios 1.4–1.9 (Rokosch and Heidecke, 2018). In contrast, the typical N/P uptake ratio for crops is about 7.5 (Xu et al., 2012).

Recognizing the limitations of direct land application of sewage sludge discussed here, Kominko et al. (2017) and Antille et al. (2017) proposed to subject the sludge to further treatment involving the addition of acid or alkali agents and mixing or coating with inorganic nutrient sources, thereby producing *organo-mineral fertilizers*. As both authors pointed out, this approach would allow simultaneously: (i) preservation of the valued organic matter content in the sludge, (ii) improvement of the main nutrient ratio (N:P:K) and (iii) a decrease in contamination. Adding alkali compounds such as lime or potassium hydroxide leads to hygienization/disinfection of the sewage sludge. Heavy metals can either be removed by acid extraction/leaching and subsequent metal precipitation, or transformed into sparingly soluble or insoluble forms by adding a natural sorbent with a strong affinity for metal ions and their hydroxides such as basaltic detritus or coal waste. A desirable nutrient ratio and availability is achieved by adding mineral fertilizers (single or multi nutrient) or their precursors, and/or acidification using a mineral acid, for example H_3PO_4. Strong acidification also destroys some pathogens. The final product is then granulated or pelletized to allow easy spreading, transport and storage. Several methods for producing organo-mineral fertilizers from sewage sludge were developed and patented in several countries, including Poland, the United States and the United Kingdom, and are described in detail in a review by Kominko et al. (2017). All methods have in common that they combine the quick release of the nutrients added by mixing or coating with mineral fertilizers with a slow release of those nutrients contained in the sludge-based core material. Greenhouse and field studies by Antille et al. (2013, 2014, 2017) suggest that such techniques allow for the production of consistent fertilizer material with a stable chemical composition designed to suit the specific nutrient requirements of crops in various soils, and to facilitate satisfactory agronomic efficiency. It is beyond the scope of this review to evaluate this approach specifically.

3.2 Inorganic contaminants

The *EU Council Directive 86/278/EEC of 12 June 1986* introduced limit values for heavy metals in sewage sludge as well as in sludge-amended soils, with the option for member states to introduce stricter limits in their national legislation. While all member states have transferred the EU limits (maximum permissible concentrations in sludge and soil) into their own regulations, only some of them, such as Germany, have made use of the option to introduce stricter limits and/or to extend the list of regulated heavy metals (Table 3).

As documented by several authors, due to the introduction of legal limit values and related source control mechanisms and environmental programmes, heavy metal contents in sewage sludge have decreased considerably since the end of the 1970s and beginning of the 1980s in most European countries as well as in other high-income countries such as the United States (Epstein, 2003; Kirchmann et al., 2017; Rokosch and Heidecke, 2018) and are usually falling safely below the legal limit values nowadays. Figure 3 illustrates this development for Germany. Given that regular documentation of heavy metal concentrations is only obligatory for the regulated elements, no long-time information is available for As and Tl that have only recently (2017) been included into the set of legal limit values for sewage sludge (Table 3).

Table 3 Limit values for total heavy metal and metalloid concentrations in sewage sludge in the EU and German legislation (AbfKlärV = Abfallklärschlammverordnung, Sewage Sludge Ordinance; DüMV = Düngemittelverordnung, Fertilizer Ordinance) in comparison to the US legislation

Metal/ metalloid	Total concentrations in mg/kg DM			
	EU directive 86/278/ EEC 1986	German AbfKlärV 1992	German AbfKlärV 2017 in conjunction with DüMV 2017	USEPA Part 503 Biosolids Rule 1993*
As	Not regulated	Not regulated	40	41–75
Cd	20–40	10	1.5	39–85
Cr	Not regulated	900	Not regulated	Not regulated
Cr(VI)	Not regulated	Not regulated	2	Not regulated
Cu	1000–1750	800	900	1500–4300
Hg	16–25	8	1	17–57
Ni	300–400	200	80	420
Pb	750–1200	900	150	300–840
Tl	Not regulated	Not regulated	1	Not regulated
Zn	2500–4000	2500	4000	2800–7500

* Land application allowed if concentration falls below the lower 'pollutant concentration limit', or if it falls between lower and upper 'ceiling concentration limit' and a cumulative pollutant loading rate limit is met by the application; USEPA Rule has additional limit values for Mo (75) and Se (100).

Consequently, the heavy metals already regulated are no longer the main focus of discussions about environmental safety of direct land application. Even though some other toxic metallic/metalloid trace elements such as Ag, Se, Sn and Ti have been recognized as potential problems in sewage sludge (Fijalkowski et al., 2017; Paz-Ferreiro et al., 2018), the focus of recent discussions and research has shifted towards organic contaminants, which will be addressed in the following section.

3.3 Organic contaminants

In recent decades, the number of organic pollutants that are released into the environment, especially from consumer products, has increased dramatically (Loos et al., 2009). A broad range of different organic contaminants enter WWTPs via domestic and industrial waste water and finally end up on the surface and in the groundwater or in the sewage sludge because conventional WWTPs are not designed specifically to remove these substances. Besides the already discussed inorganic contaminants, a range of organic contaminants can be found in sewage sludge such as organic nanoparticles, polyaromatic hydrocarbons (PAHs), polychlorinated biphenyls, perfluorinated surfactants, polycyclic musks, siloxanes, pesticides, phenols, sweeteners, personal-care products, pharmaceuticals, benzotriazoles as well as pathogens such

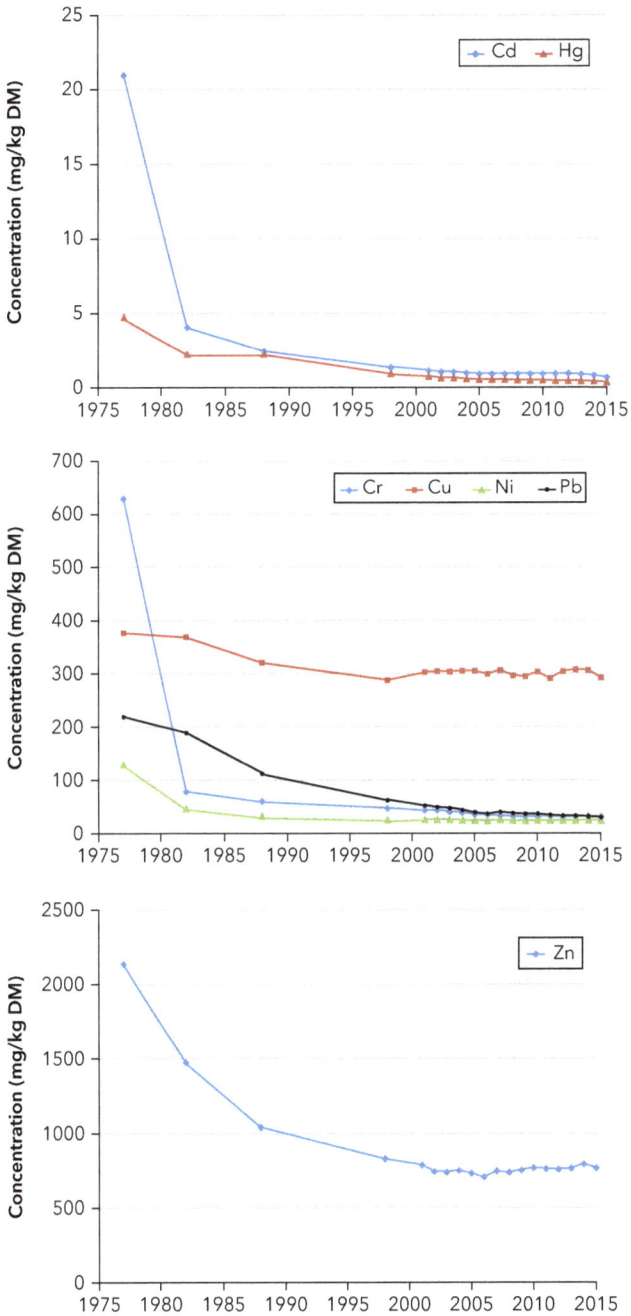

Figure 3 Changes in heavy metal concentrations in sewage sludge used for agriculture in Germany between 1977 and 2015. Source: based on data from Rokosch and Heidecke (2018).

as *Escherichia coli* O157:H7 or *Legionella* (Fijalkowsky et al., 2017). Of special relevance is the dissemination of antibiotic-resistant bacteria and the associated genes into the environment because WWTPs are considered hot spots of horizontal gene transfer (Rizzo et al., 2013). WWTPs deliver optimum growth conditions for a broad spectrum of microorganisms with respect to temperature and nutrient availability and show selective characteristics due to the presence of certain chemicals, antibiotics and heavy metals. Fornefeld and Smalla (2018) could demonstrate that after sewage sludge application, the risk of a transfer of antibiotic resistance for example by horizontal gene transfer was considerably higher than a possible contamination of plants by human pathogens that are also contained in sewage sludge.

These anthropogenic pollutants are of great concern, as they are often poorly degradable, can show high biological activity (e.g. endocrine disruptors, antibiotic-resistant bacteria) and can cause unknown additive and synergistic effects on organisms and the environment (Lindholm-Letho et al., 2017).

Many different industries, in addition to pesticide use and a broad range of personal-care products and human medications, are the origin of the multitude of organic pollutants that can be found in sewage sludge (Table 4).

More than 1500 organic compounds have been registered already, of which Harrison et al. (2006) detected and identified 516 different organic compounds in sewage sludge. This shows the main obstacle when it comes to organic pollutants: it is necessary to screen potential fertilizers made from treated waste for a broad range of complex and very different organic compounds. Moreover, these organic compounds can degrade to form intermediate or transformation products that can be of even higher concentration and toxicity than the parent compounds (Huerta-Fontela et al., 2010). Additionally, scant commercial availability of standard materials restrict the laboratory measurement of such metabolites. Even worse, some pharmaceuticals are racemic mixtures of two enantiomers of different toxicity that can hardly be distinguished by conventional analytical methods.

In Table 5, concentration ranges for some selected organic compounds that are most frequently detected in municipal sewage sludge are summarized from the literature. Pharmaceuticals and personal-care products are not listed because of the high number of different compounds belonging to these groups. The values reported indicate very high variability, which can change in relation to country and season of sampling (Lamastra et al., 2018). Especially for pharmaceuticals, a strong intra-day, inter-day or seasonal variation was observed (Petrie et al., 2015). Several drugs accumulate overnight in the urine, causing a peak concentration in the first toilet flush of the day. Differences were also detected between weekdays and weekends because some compounds that are administered in hospitals are preferentially used on weekdays. Moreover, some compounds are used seasonally such as sun blockers,

Table 4 Origin of possible organic contaminations in sewage sludge that may eventually be used for agricultural application according to Sun et al. (2011), Bergé et al. (2012), Sunday et al. (2012), Verlicchi and Zambello (2015), Abdel-Shafy and Mansour (2016) and Raju et al. (2018)

Organic compounds	Origin of the compounds
PAH - *Polycyclic aromatic hydrocarbons* 16 PAHs identified as priority pollutants, 7 of them are considered probable carcinogens	Formed during incomplete combustion of organic materials. Anthropogenic sources: residential heating, coal gasification and liquefying plants, carbon black, coal-tar pitch and asphalt production, coke and aluminium production, catalytic cracking towers and related activities in petroleum refineries as well as motor vehicle exhausts
LAS - *Linear alkylbenzene sulphonates*	Main synthetic anionic surfactants in cleaners and detergents
PFCs - *Perfluorinated chemicals (e.g. PFOS – Perfluorooctane sulfonic acid, PFOA – Perfluorooctanoic acid)*	Water, soil and stain repellents in numerous consumer products, paper and packaging, in chromium plating, electronic baths, in photographic processes, textile finishing and so forth
DEHP - *Di(2-ethylhexyl)phthalate* DBP - *Dibutylphthalate*	Plasticizers in materials such as PVC (polyvinyl chloride); hydraulic fluids, dielectric fluids and solvents
NPE - *Nonylphenolethoxylates*	Surfactants, emulsifiers, dispersants and wetting agents; used, for example, in the manufacture of textiles as surfactants and detergents
AOX - *Adsorbable organic halogen compounds* PCB - *Polychlorinated biphenyls* PCDD/F - *Polychlorinated dibenzo-p-dioxins/furans*	Solvents, solvent mixtures, oil and grease, resins, rubber, hydraulic oils, lubricants, plasticizers, disinfectants, wood protective, compound of some pesticides
OTC - *Organotin compounds*	Antifouling agents, tri-substituted OTCs are used as biocides in pesticide and fungicides, di-substituted and mono-substituted OTCs are used as catalysts and PVC heat stabilizers
PBDE - *Polybrominated diphenylethers* 209 different compounds	Flame retardants used in electrical devices, textiles, furniture, cars, planes, colours and so forth
Emerging organic contaminants	*Found in:*
PhC - *Pharmaceuticals* PCPs - *Personal-care products*	Analgesics/anti-inflammatories, anti-histaminics, hormones, antiseptics, antianginals, anti-hypertensives, hypnotics, insect repellents, antiarrhythmics, anti neoplastics, lipid regulators, UV filters, antibiotics, antiplatelet drugs, psychiatric drugs, synthetic musks, anticoagulants, antiprotozoals, contrast media, non-ionic surfactants, antidiabetics, beta-agonists, receptor antagonists, antiemetics, beta-blockers, stimulants, antifungals, diuretics
MP - *Microplastics (particles < 5 mm)*	Two major categories: 1 Primary microplastics used in many personal-care products and air-blasting media; and 2 Secondary microplastics formed through fragmentation of larger plastic items

Table 5 Concentration ranges of selected organic contaminants frequently detected in municipal sewage sludge (mean values in differently treated sewage sludge), excluding pharmaceuticals and personal-care products

Organic compounds	Range of concentration in sewage sludge (in mg/kg DM if not specified otherwise)
PAH - *Polycyclic aromatic hydrocarbons*	<0.01-16 (n = 277)[b] 0.13-28 (n = 3) reviewed by Mailler et al. (2014)
LAS - *Linear alkylbenzene sulphonates*	1300; 1500 (n = 2)[g]<1-18800 (n = 120)[b]
PFCAs - *Total perfluoralkyl carboxylates* PFOS - *Perfluorooctane sulfonic acid*	0.014-0.05 (n = 20)[d] 0.015-0.6 (n = 20)[d]
DEHP and DBP - *Di(2-ethylhexyl) phthalate and dibutylphthalate*	0.4-149 (median 11; n = 130)[i] <0.02-3514 (mean 58; n = 306)[c]
ΣNP and NPnEOs - *Nonylphenol and Nonylphenolethoxylates*	NP: 0.02-2530 (mean 128; n = 24)[e] NP1EO: 0.15-850 (mean 40; n = 18)[e] 0.4-103 (median 6.2; n = 130)[i]
AOX - *Adsorbable organic halides*	170; 260 (n = 2)[g] 350[h] 130-2000 (median 275; n = 16)[a]
PCB - *Polychlorinated biphenyls*	0.04; 1.4 (n = 2)[g] 0.0006-7 (n = 86)[b] <0.006-1.93 (n = 3) reviewed by Mailler et al. (2014)
PCDD/F - *Polychlorinated dibenzo-p-dioxins/furans*	0.000035[h] 3.0-69 ng/kg DM (n = 59)[b]
OTC - *Organotin compounds*	<0.02-9 (mean 0.63; n = 90) reviewed by Clarke and Smith (2011)
PBDE - *Polybrominated diphenylethers*	0.158-9.4 (mean 2.8; n = 8)[f] 0.005-4.7 (mean 1.4; n = 123)[c]

[a] Frost et al. (1993), [b] JRC (2001), [c] Clarke and Smith (2011), [d] Sun et al. (2011), [e] Bergé et al. (2012), [f] Cincinelli et al. (2012), [g] Kapanen et al. (2013), [h] Wiechmann et al. (2013), [i] Lamastra et al. (2018).

preparations against hay fever or drugs against common cold, resulting in high seasonal variability of these compounds in waste water. Additional variations in the amounts of waste water occur in tourist regions.

Inappropriate sampling strategies are considered the largest weakness of reported data, causing a lack of understanding of the spatial and temporal variations of emerging contaminants. According to Petrie et al. (2015), current approaches tend to use discrete grab sampling at low inter-day frequency and often without intra-day repetition. This approach only yields a snapshot of contamination for a specific point in time.

A review by Verlicchi and Zambello (2015) on the influence of different sewage sludge treatment processes on the concentrations of 152 pharmaceuticals and 17 personal-care products revealed that most compounds were reduced or attenuated during sludge treatment (digestion, composting, conditioning, drying etc.), but some groups of compounds (analgesics, antibiotics, hormones, antiseptics) were enriched in comparison to the primary sludge. Anaerobic digestion and composting of sewage sludge reduced the concentrations of most investigated organic compounds (Verlicchi and Zambello, 2015) and pharmaceuticals by around 30% (Malmborg and Magnér, 2015). Composting of sewage sludge resulted in a reduction of concentrations of LAS, NPE/NPs and PAH by 84%, 61% and 56%, respectively, after 26 days in a pilot study (Kapanen et al., 2013).

A range of factors that cause high variability in the contamination concentrations clearly suggest that it is hardly possible to compare results of different studies and provide mean concentrations of emerging contaminants in sewage sludge. Nevertheless, some general trends were observed. The concentrations in sewage sludge of some important pollutants such as polychlorinated dibenzo-p-dioxins and dibenzofurans (PCDD/F), polychlorinated biphenyls and AOX, which were already in the focus of research in the 1990s, decreased significantly over the last few decades due to changes in legislation (Zennegg et al., 2013; Fijalkowsky et al., 2017). This underlines the importance of legal limit values for critical compounds in sewage sludge. Out of 152 pharmaceuticals and 17 personal-care products found in sludge-amended soils, Verlicchi and Zambello (2015) identified estradiol, ciprofloxacin, ofloxacin, tetracycline, caffeine, triclosan and triclocarban as the most critical compounds because their risk quotient, which is the ratio between pollutant concentration and its predicted no-effect concentration, was ≥1.

An environmental risk assessment of the concentration data in Table 5 is only possible in relation to toxicity tests. Many studies were undertaken to characterize the toxicity of chemicals found in sewage sludge for different environmental components (e.g. JRC, 2001; Kraus and Seis, 2015) or the predicted long-term accumulation of toxic compounds in soil (Weissengruber et al., 2018). Very often, these studies focus only on selected compounds, missing out on the complete picture.

It is difficult to draw conclusions about the eco-toxicity of a particular sludge solely from concentrations of organic contaminants because these compounds do not appear individually but as a complex mixture, which could lead to unwanted synergistic effects (e.g. Seiler and Berendonk, 2012). Moreover, degradation of chemicals can result in partly unknown degradation products, and a range of factors can change the toxic character of the compounds, including site-specific parameters. The environmental ubiquity of several potentially toxic organic compounds underpins the need to improve the knowledge on their occurrence, environmental fate and ecological impacts (Petrie et al., 2015). As mentioned before, the legally regulated limit values for organic contaminants are important to reduce contaminations in the long-term, but they do not guarantee environmentally safe fertilizers, as the list of emerging toxic compounds with unknown toxicity properties is long.

To get a deeper insight into the multitude of organic contaminants in sewage sludge, many review articles are available, but only a few are mentioned here. Numerous reviews deal with the contamination of sewage sludge by pharmaceuticals, personal-care products and other emerging organic pollutants (e.g. Harrison et al., 2006; Clarke and Smith, 2011; Jones et al., 2014; Mailler et al., 2014; Petrie et al., 2015; Lindholm-Letho et al., 2017). Treatment options affecting the degradation of such compounds are also under evaluation (e.g. Mailler et al., 2014; Ahmed et al., 2017; Bloem et al., 2017; Grandclément et al., 2017; Alvarino et al., 2018). Moreover, complex topics such as the contamination of sewage sludge by nanoparticles, plastics, pathogens or antibiotic resistance are discussed comprehensively (e.g. Kunhikrishnan et al., 2015; Chen et al., 2016; Nnadozie et al., 2017; Karbalaei et al., 2018; Raju et al., 2018; Rezania et al., 2018).

In conclusion, the current organic contamination of sewage sludge is of concern due to possible ecological impacts (e.g. endocrine disruption, dissemination of antibiotic-resistant bacteria) with negative effects for flora and fauna as well as for human health. The insufficient knowledge about the environmental fate of these compounds indicates that integrated analytical approaches are necessary to measure the ecological impact of sewage sludge application. Hence, today direct land application of sewage sludge without further processing to decrease organic contamination (e.g. by incineration) does not seem recommendable from our point of view although valuable nutrients and organic substances are partly lost by processing.

4 Products derived from the treatment of waste water and further (post-treatment) processing

4.1 Technical processes involved in treatment of waste water

Driven by discussions about potential risks of direct land application, but also by purely operational needs such as the prevention of undesired scaling of pipes and pumps in WWTPs by spontaneous precipitation of inorganic salts, there has been intensive research on developing alternative strategies to use valuable nutrients, especially P, contained in waste water and sewage sludge. Several recovery routes were explored; they are illustrated together with examples of related products in Fig. 4. A detailed description of the recovery processes currently available or under investigation can be found in the recent review on this topic by Kratz et al. (2019).

4.1.1 Precipitation of N and P salts from waste water

Recovery of N and P from waste water may already take place on site, that is in the WWTP, by precipitating nutrients from the waste water stream. The most

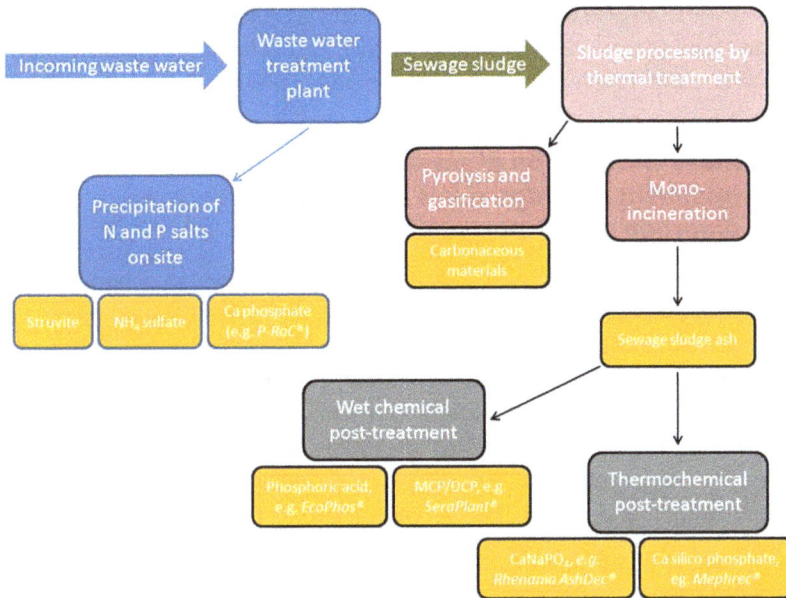

Figure 4 Simplified diagram of the options for nutrient recovery from waste water treatment and the examples of related products (denoted by ®). P-RoC® = precipitated salt from P-RoC® process (Phosphor Recovery by Crystallization) [Ehbrecht et al. (2016)].

prominent product resulting from this recovery route is $NH_4MgPO_4 \times 6H_2O$, known as struvite. Struvite can either be obtained by direct precipitation from a P-rich side stream of the WWTP resulting from the stimulated anaerobic P release, or after forced digestion of the biomass by hydrothermal and/ or chemical pre-treatment (Adam, 2018). The first two options only work for EBPR facilities, whereas the chemical pre-treatment, for example addition of sulphuric acid, also allows to precipitate struvite from sewage sludge containing chemically precipitated P, thereby increasing the P recovery rate from less than 25% up to around 50% of the P input to the WWTP (Antakyali et al., 2013; Egle et al., 2015).

Alternatively, P may be precipitated on site by dosing of lime, calcium chloride or calcium silicate hydrate, resulting in pure calcium phosphates such as brushite (dicalcium phosphate) or hydroxyapatite, or a mix of struvite and calcium phosphate (Ehbrecht et al., 2016; Bogner and Ortwein, 2018).

An option to recover N on site at the WWTP is ammonia air-stripping from the nutrient-rich side stream generated by dewatering of the sewage sludge, followed by recovery of ammonia with sulphuric acid (Abwasserverband Braunschweig, 2014; Boehler et al., 2015). The resulting product is diammonium sulfate ($(NH_4)_2SO_4$), which has been known as a mineral fertilizer of high plant availability for many years.

Precipitation of N and P salts has been successfully applied for other waste streams (apart from municipal waste water), for example for the treatment of dairy waste waters (Ahmed et al., 2018) and animal slurry (Brown et al., 2018).

4.1.2 Mono-incineration of sewage sludge and post-treatment of sludge ashes (acid leaching/ digestion or thermochemical treatment)

Mono-incineration of sewage sludge is well established on a large scale in some European countries (see Section 3.1). While N and organic matter are removed by incineration, the resulting sewage sludge ashes (SSA) contain up to 13% P (Krüger et al., 2014). The P recovery potential is very high, with >90% of the total P load of a WWTP (Krüger and Adam, 2015).

The loss of mass resulting from incineration not only leads to concentrating mineral elements such as P, but also to an enrichment of non-volatile heavy metals (Krüger and Adam, 2014). Depending on the quality of the input sludge, heavy metal contents in the resulting SSA may vary considerably (see Section 4.3, Table 7). Even though low heavy metal contents may allow direct use of untreated SSA as a fertilizer, the plant availability of nutrients from this material is rather limited; for example, at high temperatures, P is transformed into stable, sparingly soluble or insoluble calcium phosphates such as hydroxyapatite or whitlockite (Table 6; reviewed by Kratz et al., 2018). Therefore, various post-treatment approaches were developed in the past to derive a plant-available P fertilizer from the SSA that is low in heavy metals. The two general strategies for post-treatments are acid digestion or leaching, and thermochemical treatment. The following description is based on Adam (2018) and Kratz et al. (2019).

Acid digestion or leaching has been carried out with sulphuric, nitric, chloric and phosphoric acid and is already practiced at industrial scale in order to convert P into a water-soluble form. The fertilizer industries tested part-substitution of phosphate rock by SSAs in their conventional production processes based on *acid digestion*. However, high Fe and Al contents in SSA pose operational as well as product-quality problems. In contrast to digestion, leaching procedures aim to separate P from the source material. Phosphorus and metals are dissolved simultaneously, followed by separation of P from metals by ion exchange, solvent extraction, selective precipitation, electro dialysis or membrane filtration. Phosphoric acid is the product that is in the focus, known for example under the trade names EcoPhos® or TetraPhos®. A previous leaching technique known under the trade name LeachPhos® is no longer available, but still mentioned because samples of the resulting material were used in a number of publications cited here. Recovered phosphoric acid can be used to produce fully water-soluble superphosphates or water-insoluble, but plant-available dicalcium phosphates for fertilizer applications.

Different processes were developed for *thermochemical post-treatment* of SSAs. The Mephrec® process is based on gasification of SSA and/or sewage sludge in a cupola furnace together with additives at high temperatures of up to 2000°C. Volatile heavy metals are transferred into the gas phase and are removed in the flue gas treatment system. Heavy metals with high boiling points are collected in a separate metal phase. Phosphorus is intended to be transferred into a CaO-SiO_2 slag, forming plant-available silico phosphates that are well known from 'Thomas slag', a fertilizer product from steel industries until the late 1980s. However, recent studies showed that P is transferred to all three major output flows (metal, slag and filter dust from gas cleaning) (Hagspiel, 2018).

The same type of calcium silico phosphates can be produced by introducing SSA into hot blast (steel mill) furnace slag. The calcium silicates of the steel making slag react at temperatures of approx. 1600°C with the phosphates present in the ash (Rex et al., 2014). Another process known under the trade name AshDec® intends to remove volatile heavy metals via the gas phase at approx. 1000°C and at the same time transform P present in SSA into plant-available mineral compounds. Different additives were investigated in terms of heavy metal removal and plant availability of the resulting phosphates. Utilization of the additives $MgCl_2$ and $CaCl_2$ led to extensive removal of heavy metals including Cu and Zn (Adam et al., 2009). However, due to an unsatisfactory fertilization effectiveness of the resulting materials on neutral and alkaline soils, another approach (Rhenania AshDec) was tested, involving a thermochemical treatment of SSA under reducing conditions with alkaline compounds such as Na_2SO_4 or Na_2CO_3. The plant-available phosphate species $CaNaPO_4$ is formed (Table 6), known as 'Rhenania Phosphate' produced from phosphate rock and soda in the last century (Herzel et al., 2016).

Mono-incineration with its related post-treatments can also be a viable option to make use of other waste streams such as slaughtering wastes (Cascarosa et al., 2012).

4.1.3 Pyrolysis and gasification

Besides the production of ashes in an oxygen-rich atmosphere, there are also efforts to treat organic materials by thermochemical conversion in an oxygen-limited environment to produce carbonaceous materials by gasification or dry and wet pyrolysis. While the main focus of research was on the mitigation of greenhouse gas (GHG) emissions by the production of green fuels (Jensen et al., 1998; Demirbaş, 2001; Day et al., 2005; Gaunt and Lehmann, 2008) with carbonaceous materials as by-products, this intention has changed in the last decade towards the application of specifically produced carbonaceous materials as soil conditioners or fertilizers. These materials are often summarized using the term 'biochar' and are of highly varying characteristics caused by the differences in feedstock as well as peak pyrolysis temperature, pressure and retention time during production (Neves et al., 2011; Meier et al., 2013; Quicker and Weber, 2016; Sun et al., 2017).

The interactions between feedstock biomass and temperature are not fully understood (Aller, 2016). The type of feedstock mainly influences the nutrient and possible contaminant concentrations in the final carbonaceous materials. While carbonaceous materials of nutrient-rich feedstock such as animal manure, animal bones or sewage sludge contain considerable amounts of nutrients, herbaceous or woody feedstock result in material with low-nutrient content (Tsai et al., 2012; Xu et al., 2014; Zielińska et al., 2015; Laghari et al., 2016). Additionally, peak pyrolysis temperature influences the final nutrient content in the carbonaceous material (Crombie et al., 2015; Bruun et al., 2017). Whereas P volatilization does not start below temperatures of 1200°C (Zhang et al., 2012), volatilization of N occurs at much lower temperatures resulting in N losses of up to 73% during pyrolysis (Gaskin et al., 2008; Aller, 2016). Because most processes aim to produce stable carbonaceous materials to mitigate GHG emissions and sequester carbon in soils, nutrients are often not easily plant available (Angst and Sohi, 2013; Wang et al., 2012; Gul and Whalen, 2016; Robinson et al., 2018). Dai et al. (2016) summarized the pyrolysis effects on the solubility of P in carbonaceous materials, and Yuan et al. (2016) reported about the effects of peak temperature on N, P and K availability in carbonaceous materials derived from sewage sludge. The influence of the above-mentioned processing parameters impedes easy evaluation of the fertilizing potential and the speciation of P in carbonaceous materials produced, for example from sewage sludge (Hossain et al., 2011; Song et al., 2014; Huang and Tang, 2016; Robinson et al., 2018; Sun et al., 2018). Gonzaga et al. (2017) demonstrated that two different pyrolysis methods resulted in contradicting effects on germination rate and early shoot and root growth of corn. While material from a 'Top-Lit Updraft retort unit' (550–700°C, 3 h) reduced the germination rate of corn seeds to nearly zero, the material from a muffle furnace (600°C, 1 h) increased the germination rate (Gonzaga et al., 2017).

Highly controversial and inconclusive discussions about the fertilizing potential of carbonaceous material have been ongoing for about 15 years. Case study results show the soil- and climate-dependent fertilizing effects (Crane-Droesch et al., 2013; Macdonald et al., 2014; Butnan et al., 2015; Archontoulis et al., 2016; Agegnehu et al., 2017; Subedi et al., 2017) as well as the effects caused by particle size of the carbonaceous materials used (Jaafar et al., 2015; He et al., 2018). It is beyond this chapter to provide a full overview of these results and discussions. Interested readers are referred to the review articles

(Jeffery et al., 2011, 2015; Ibarrola et al., 2012; Spokas et al., 2012; Biedermann and Harpole, 2013; Aller, 2016; Gul and Whalen, 2016; Agegnehu et al., 2017; Hussain et al., 2017; Nguyen et al., 2017) that provide good information about the topic.

4.2 Plant availability of processed products

As mentioned before, the plant availability of a fertilizer depends on the binding forms of the nutrients it contains. Table 6 lists the P binding forms in recycled products derived from waste water or sewage sludge, which were identified in the literature.

In order to assess the agronomic efficiency of the products, vegetation trials must be performed. Kratz et al. (2018, 2019) reviewed a large number of vegetation trials that investigated fertilizing effects of recycled products derived from sewage sludge. Two general limitations were noted regarding the validity of conclusions drawn from such experiments. First, most of the tested recycled products were not yet generated at an industrial scale, but originated from technical trials at the laboratory or pilot scale. Therefore, product characteristics and quality may vary considerably depending on the origin of the samples used for testing. Secondly, the vegetation trials reviewed showed a broad variability in their experimental design. Obviously, solubility and plant uptake of P, as well as of any other essential plant nutrient, do not only depend on the P speciation of a given fertilizing material, but of course also on the complex interactions taking place at the fertilizer/soil(water)/plant(root) interface. In other words, soil and plant characteristics influence results of vegetation studies just as much as fertilizer properties. Accordingly, the data on fertilizing effects of different recycled products cover a broad range, as illustrated in Fig. 5. Fertilizing effects were evaluated based on P uptake in most of the studies reviewed by Kratz et al. (2018, 2019). For that purpose, P uptake of the test fertilizer was related to that of a highly water-soluble mineral reference or control fertilizer such as triple superphosphate. This was then referred to as relative fertilizer efficiency (e.g. Cabeza et al., 2011; Lemming et al., 2017b), relative P fertilizer effectiveness or uptake (e.g. Möller et al., 2018; Vogel et al., 2015) or MFE (e.g. Delin, 2016; Øgaard and Brod, 2016). In order to avoid further confusion of terminology, we refer to it as relative P uptake (rPU) here. Following the majority of the studies reviewed by Kratz et al. (2018, 2019), we calculated a net value, i.e. P uptake (PU) of the zero-P treatment was subtracted from each of the other treatments (including the treatment with the reference fertilizer) according to the following equation:

*Net rPU(%) = (PU test fertilizer - PU zero P) * 100/PU control fertilizer - PU zero P*

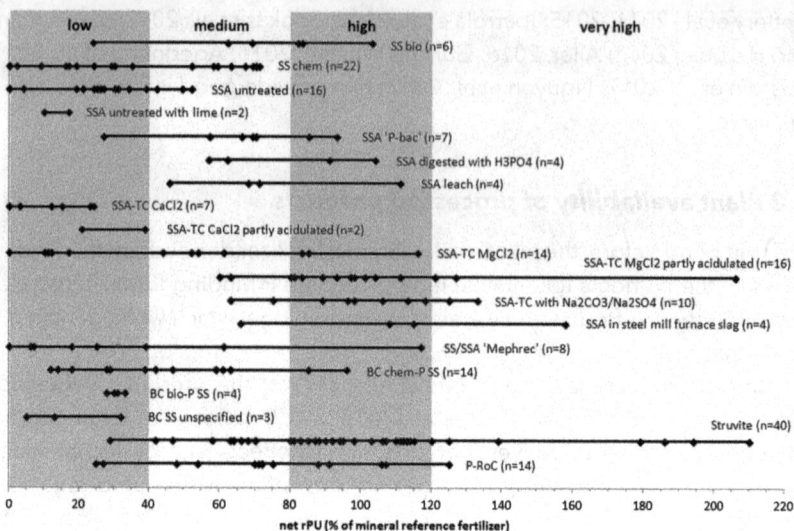

Figure 5 Net relative P uptake (net rPU) of different fertilizers based on recycling from waste water treatment (in % of water-soluble reference fertilizer). Abbreviations: *SS* sewage sludge (biological or chemical P removal), *SSA* sewage sludge ash, *P-bac* treatment involving bacteria, *leach* treatment involving acid leaching, *TC* thermochemically treated (with different additives), *BC* biochar/carbonaceous material, *P-RoC* precipitated salt from P-RoC® process (see text for further explanations). Source: Kratz et al. (2019).

Even though the data do not allow general conclusions yet, some trends can be observed. Sewage sludge treated only by BPR tends to have higher rPU than chemically precipitated sludge, and may even reach the same order of magnitude as a water-soluble mineral fertilizer. This can be explained by the P forms resulting from the different types of treatment (see Section 4.1 and Table 6).

Recovery of N and P salts from waste water is already practiced at an industrial scale by a large number of WWTPs. High rPU of the resulting products, mainly struvite or a mix of poorly crystallized Ca-deficient hydroxyapatite with struvite (P-RoC® process), has been demonstrated many times; however, different experiments display a wide range of values (Fig. 5; with data from Römer, 2006; Kratz et al., 2010; Cabeza et al., 2011; Wilken and Kabbe, 2015; Zeggel et al., 2015; Vaneeckhaute et al., 2016; Duboc et al., 2017; Lemming et al., 2017; Meyer et al., 2018; Vogel et al., 2015; Wollmann et al., 2018). The large variation in rPU of struvite may be related to different product characteristics, such as varied crystal size and structure, as well as impurities in the crystals (Cabeza et al., 2011; Möller et al., 2018). Other possible reasons are associated with the experimental design of the vegetation trials, including variations in particle size of the products tested, for example after sieving or milling, which will affect their dissolution rate (Degryse et al., 2017; Lemming et al., 2017). Furthermore, differences in parameters, such as test plant species, type and quality of test substrate (e.g. pH, texture, organic matter content, P status) and/or duration of growth experiments, may play an important role (Kratz et al., 2018, 2019). Pure struvite has been shown to achieve good fertilizing performance on acidic as well as neutral or slightly alkaline soils, whereas products from the P-RoC® process tend to perform better on (moderately) acidic than on neutral or alkaline substrates (Cabeza et al., 2011; Zeggel et al., 2015; Ehbrecht et al., 2016; Meyer et al., 2018; Wollmann et al., 2018). This may be explained by a large proportion of apatite-like P compounds in the latter product, whose dissolution requires an acidic soil environment (Meyer et al., 2018).

Mono-incineration generally decreases the rPU of the resulting SSA because it leads to the formation of stable Ca phosphates (Table 6). Thermochemical treatment at high temperatures (approx. 1000°C) was more successful with $MgCl_2$ than $CaCl_2$, producing Mg-containing P forms that proved to be plant available (stanfieldite, farringtonite, see Table 6), but only in an acidic soil environment, or when the product had been partially acidulated. Products from thermochemical treatment with Na_2CO_3/Na_2SO_4 performed satisfactorily in vegetation trials on neutral or alkaline soils (Schick, 2009; Kratz et al., 2010; Nanzer et al., 2014; Severin et al., 2014; Wilken and Kabbe, 2015; Vogel et al., 2015; Zeggel et al., 2015; Lemming et al., 2017; Meyer et al., 2018).

Acid digestion or leaching (also in combination with ion exchange treatment) of SSA may result in products with high rPU because plant-available P forms such as dicalcium or monocalcium phosphate are generated (Table 6; Wilken and Kabbe, 2015; Zeggel et al., 2015; Kratz et al., 2017). Zeggel et al. (2015) tested products based on biological (in situ) digestion of SSA by addition of bacteria, achieving promising results (referred to as P-bac in Fig. 5). However, large differences in chemical solubility and agricultural performance of the two supposedly similar products tested by the authors illustrate the difficulties in evaluating products/techniques that are still in an early developmental stage.

Test products from blowing SSA into fluid steel mill slag resulted in good P availability (Severin et al., 2014; Lekfeldt et al., 2016).

For test products from the high temperature Mephrec® process, quite contradictory results were reported by Cabeza et al. (2011), Wilken and Kabbe (2015), Zeggel et al. (2015) and Wollmann et al. (2018). The performance of the slag appears to depend on acidity/alkalinity of the test substrate, but no clear picture can be drawn from the data currently available. Test products used in growth trials so far stem from very different production conditions, ranging from the lab scale to demonstration/pilot scale, which probably explains their variable performance.

Table 6 P forms in recycled products derived from waste water or sewage sludge according to literature data

Product	P forms
Precipitated salts	
Struvite	$NH_4MgPO_4 \times 6H_2O$ may contain impurities or occur as a mix with other P-Mg-compounds, for example NH_4MgPO_4/ $MgHPO_4$
P-RoC®	HAp-like CaP phases of poorly structured small crystals (Ca-deficient HAp), mixed with struvite
Sewage sludge ashes (SSA) and derived products	
SSA untreated	TCP/whitlockite HAp $Ca_9Al(PO_4)_7$ Al/Fe phosphate
SSA thermochemically treated with $CaCl_2$ (1000°C)	ClAp HAp Al phosphate
SSA thermochemically treated with $MgCl_2$ (1000°C)	ClAp $Ca_4Mg_5(PO_4)_6$ (stanfieldite) $Mg_3(PO_4)_2$ (farringtonite)
SSA thermochemically treated with Na_2CO_3/Na_2SO_4 (1000°C) (RhenaniaAshDec®)	$CaNaPO_4$ $CaKPO_4$ $Ca(Na,K)PO_4$ $Ca_{13}Mg_5Na_{18}(PO_4)_{18}$
Products from acid/alkaline leaching of SSA (e.g. LeachPhos®, EcoPhos®)	(amorphous) (Al-)Ca phosphate Al phosphate
Products from acid treatment of SSA (e.g. SeraPlant®/earlier name: RecoPhos®)	MCP/DCP Al/Fe phosphate
Carbonaceous materials (Pyrolysis products)	
- From Bio-P (EBPR) sewage sludge	Orthophosphates (i.e. PO_4 in various types of phosphate salts or associated with other mineral phases) Polyphosphates Pyrophosphates
- From Bio-P (EBPR) sewage sludge, post-treated with $Al_2(SO_4)_3$	Short-chain polyphosphates Al phosphate
- From sewage sludge chemically precipitated with $FeCl_2$	TCP/whitlockite

Abbreviations for the Ca phosphates: Ap = apatite, HAp = hydroxyapatite, ClAp = chloroapatite ($Ca_5(PO_4)_3Cl$), OCP = octacalcium phosphate ($Ca_8H_2(PO_4)_6$*$5H_2O$), TCP = tricalcium phosphate (occurs mostly as whitlockite in natural materials: $Ca_{3x}(Mg,Fe^{2+})_x(PO_4)_2$), DCP = dicalcium phosphate (may occur as anhydrous or hydrous form = monetite ($CaHPO_4$) or brushite ($CaHPO_4$*$2H_2O$)), MCP = monocalcium phosphate.
Source: modified from Kratz et al. (2018, 2019).

Pyrolysing sewage sludge at temperatures between 400°C and 500°C has not been successful in producing a highly plant-available P fertilizer yet. Positive results reported from a trial by Kern et al. (2016) could not be replicated by other researchers (Appel et al., 2016; Appel and Friedrich, 2017; Steckenmesser et al., 2017; Wollmann et al., 2018). The only promising pyrolysis products could be those from low-temperature pyrolysis of Bio-P (EBPR) sewage sludge: long-chain polyphosphates contained in EBPR sludge are converted to di-, tri-, meta- and short-chain polyphosphates during pyrolysis at a temperature below 400°C (Steckenmesser et al., 2017). These short-chain polyphosphates have high plant availability (Torres-Dorante et al., 2005). However, no data on vegetation trials with such products was available to the authors at the time of writing this review.

The highly variable results presented here illustrate clearly that standardized experimental conditions are an indisputable prerequisite in order to reach a viable conclusion regarding the plant availability of different types of sludge as well as sludge-derived fertilizer products.

4.3 Inorganic contaminants in processed products

Processing of sewage sludge to generate fertilizing products involves treatment steps such as precipitation or incineration, leading to changes in concentrations of mineral elements, including heavy metals. Given that many of the respective products are not yet generated at an industrial scale, the data on heavy metal concentrations are scarce and often limited to very few samples, which may not even be representative, because the technical development is still ongoing. Therefore, the following examples from Germany (Table 7), as well as any conclusions based on them, must be understood as preliminary.

The *German Fertilizer Ordinance (DüMV)* provides the limit values for As, Cd, Cr, Hg, Ni, Pb and Tl in fertilizers. Copper (Cu) and Zn, which were also limited in earlier versions of the *DüMV*, have been excluded from regulation a few years ago, based on the argument that they should be viewed as essential micronutrients rather than contaminants. However, the difference between nutritional value and contamination in this case is simply one of elemental concentration or dosage, so they are still discussed as potentially toxic elements in this review. Recently, the regulation of uranium (U) has also been considered because this element is known to occur in high concentrations in phosphate rock, the raw material used to produce mineral P fertilizers, and is characterized by high toxicity (De Kok and Schnug, 2008). It is therefore included in this overview as well.

As a first evaluation step, the concentration data can be compared to the legal limit values (here: *German Fertilizer Ordinance* in conjunction with *Waste*

Table 7 Ranges of P and heavy metal concentrations in unprocessed sewage sludge and selected products generated by post-treatment of sewage sludge in Germany in comparison to a commercial phosphate rock-based triple superphosphate

Product	P (%)	Heavy metals and metalloids (mg/kg DM)									
		As	Cd	Cr	Cu	Hg	Ni	Pb	Tl	U	Zn
Unprocessed sewage sludge[a,f,g]	1-4	2-8	0.6-1.1	18-72	195-783	0.4-2.1	18-43	20-57	<0.02-0.2	1.0-1.6	733-1236
Precipitated salts											
Struvite (n = 11)[a,c,d,g]	8.7-18	0.3-2.7	<0.01-0.3	1.5-16	2.6-132	0.06-0.6	<0.5-19	<0.1-13	<0.04	0.14-0.2	15-258
Ca phosphate P-RoC® (n = 2)[c,g]	5.8-8.4	0.6-0.9	<0.01-0.5-	11-17	<1.0-6.6	<0.01-0.1	<2.7-10	<0.1-11	<0.01	1.1/0.3	<5
Mono-incineration ash (n = 43)[b,c]	1.4-12	5.9-54	0.1-12	0.1-989	114-3467	<0.01-2.6	42-424	9-404	<0.01-0.4	1.4-26	867-4076
Wet chemical post-treatment											
LeachPhos® (acid leaching) (n = 1)[d]	13	10	3.8	34	851	0.2	14	25	nd	nd	1390
RecoPhos® (acid digestion) (n = 1)[c]	16	11	1.8	71	464	0.1	47	75	0.3	5.5	1434
Thermochemical treatment											
Rhenania AshDec® (n = 2)[d,g]	7.7	4.0/6.8	0.4/0.9	127/71	601/371	0.7/0.4	56/41	60/31	nd/0.3	nd/7.6	1737/1633
Mephrec® slag (n = 2)[c,d]	4.4/5.5	4.7/2.6	0.3/<0.01	110/60	115/76	0.7/<0.01	17/5.3	4.2/0.5	nd/<0.01	nd/9.4	85/70
Pyreg® biochar (n = 2)[e]	5.4/5.6	nd	1.3/0.6	105/126	952/872	nd	76/77	67/66	nd	nd	2049/1995
Triple super phosphate (n = 9)[a,d,g]	19-23	3.7-15	3.8-35	79-288	4.7-37	0.04-0.2	12-43	1.7-15	0.2-0.5	58-176	178-603
German limit values[*]		40	1.5[**]	No limit value	900	1	80	150	1	No limit value	4000
German background concentrations for agricultural topsoils (minimum value for most sensitive soil category)[***]		2.2	0.1	7.8	6.0	0.03	2.8	13	0.05	0.3	23

[*] DüMV in conjunction with AbfKlärV (see Section 3.2, Table 3).

[**] Cd limit is defined in relation to dry substance if fertilizer has a P content of <5% P_2O_5, for fertilizers >5% P_2O_5, it is 50 mg Cd/kg P_2O_5.

[***] LABO (2017); nd = not determined.

References: [a]Analyses by the chapter authors, [b]Krüger and Adam (2014), [c]Zeggel et al. (2015), [d]Kraus and Seis (2015), [e]Appel and Friedrich (2017), [f]Rokosch and Heidecke (2018) averages for 2015, [g]Koch et al. (2018).

Sewage Sludge Ordinance), or to background concentrations for agricultural topsoils (for Germany: LABO, 2017). The term 'background concentration' refers to the geogenic concentration of an element or substance found in a soil and is defined by its parent material, including geogenic transformations and pedogenic redistribution of substances, plus ubiquitous distribution of substances caused by diffuse inputs into the soil (LABO, 2017). Background concentrations vary considerably depending on the soil parent material. For a risk assessment, it appears sensible to work with the minimum value (i.e. the most sensitive soil category) in order to create a sort of 'worst-case scenario'. As shown in Table 7, the mean heavy metal concentrations in products made from sewage sludge are often below the current legal limit values (no such limit exists for U yet). However, in many cases, they exceed the background concentrations for agricultural topsoils. This implies that in the long term, accumulation of heavy metals must be expected as a result of fertilization with sewage sludge and the related products.

In order to better assess this expected accumulation, heavy metal loads, that is the amounts of heavy metals applied to the field in relation to a given nutrient dosage (in this case P), can be calculated. Various recycled fertilizers show different P-to-heavy metal ratios, resulting in different heavy metal loads associated with fertilization. Full risk assessments for sewage sludge-based fertilizers were published recently by Kraus and Seis (2015), Kraus et al. (2018, 2019) and Weissengruber et al. (2018). They carried out the comparative assessment of product quality based on the P-related contaminant load. As a reference, they used a conventional water-soluble mineral P fertilizer and assumed uniform fertilization rate adjusted to crop needs. In comparison to the mineral reference fertilizer, sewage sludge and derived products were associated with considerably lower loads of Cd and U. In contrast, raw and thermochemically treated SSAs led to higher loads of Cu and Zn than the mineral reference, whereas Cu and Zn loads from the acid-leaching products (EcoPhos®, TetraPhos®) and struvite were in the same order of magnitude or even lower than those from the mineral fertilizer.

To evaluate the risk emanating from the calculated heavy metal loads, a mass balance approach was followed, that is additional input routes for potentially toxic elements such as atmospheric deposition and liming were considered, and the sum of inputs was balanced with outputs by plant offtake and via leaching losses (Kraus and Seis, 2015; Kraus et al., 2018, 2019; Weissengruber et al., 2018). Regarding heavy metal loads, atmospheric deposition of heavy metals played a more important role than fertilizer-bound inputs in the case of struvite. In contrast, products derived from post-treatment of sewage sludge often resulted in inputs of Cr and Cu (in some cases also As, Hg and Zn) that considerably exceeded the input of these elements by atmospheric deposition (Kraus and Seis, 2015; Weissengruber et al., 2018). The authors of both studies agreed that looking at a period of 100 (or 200) years, no unacceptable risk would arise from the use of P recycled fertilizers in terms of human health, soil protection or groundwater contamination regarding heavy metal inputs. Nevertheless, they emphasized that the actual accumulation risk depends largely on the specific soil and climatic conditions of a location. In some cases, Zn was identified as a potential long-term risk to soil organisms or groundwater; however, further studies are required to better qualify this risk.

As can be seen in the *European Sewage Sludge Directive* dating from 1986, the EU has acknowledged quite early that concentration limits do not guarantee environmental protection, and proposed maximum permissible heavy metal loads per hectare and year. Unfortunately, even around 30 years later, only few member states have implemented that option in their national legislation (see Mininni et al., 2015 for detailed comparison of limit values across EU member states). The development and implementation of environmentally sustainable legislation for this subject remains a future challenge.

4.4 Organic contaminants in processed products

Struvite is a precipitation product that can be obtained in WWTPs or during food processing, and can contain a certain proportion of the original contamination of the parent material. However, generally, the contamination with organic contaminants is low because more than 98% of the contaminants stay in the solution when the process is optimized (Ronteltap et al., 2007). Therefore, precipitation products generally show lower organic contamination than the original material.

It is assumed that thermal treatment of organic materials such as sewage sludge at high temperatures (850-950°C) destroys organic contaminants and pathogens to a large extent (Wiechmann et al., 2013; Lindberg et al., 2015). Most likely, this is the reason why very few studies have investigated the organic contamination of fertilizers produced from thermochemically treated SSA (Table 8).

During the production of carbonaceous materials, much lower temperatures are applied. Therefore, some organic contaminants may remain intact. However, pharmaceuticals such as triclocarban and triclosan and the surfactant nonylphenol were almost completely destroyed in carbonaceous materials produced at 500°C (Ross et al., 2016). Pyrolysis of sewage sludge at 400°C for 1 h diminished the estrogenicity of the sample by 95% (Hoffmann et al., 2016). Nevertheless, during pyrolysis at temperatures below 620°C the formation of polycyclic aromatic hydrocarbons (PAH) was observed (De la Rosa et al., 2016). The authors determined PAH ($\sum^{16}$PAHs) contaminations of 959 ± 62 µg/kg DM in materials produced from sewage sludge and even higher concentration of 2613 ± 1380 µg/kg DM in materials produced from pinewood. The PAH contamination in carbonaceous materials is highly variable depending on different parent materials used as well as varying process parameters such as temperature and duration of the process. However, contradictory results were found in literature. Pyrolysis of sewage sludge at 430°C for 120 minutes caused a significant decrease in the total PAH concentration in comparison to that in feedstock: the PAH concentration was reduced 9.5- to 10.9-fold in municipal sewage sludge and 10.1- to 11.8-fold in industrial sludge (Fristak et al., 2018). Luo et al. (2014) investigated the PAH concentration in carbonaceous materials produced from sewage sludge at different pyrolysis temperatures and found decreased PAH concentrations from 200 to 400°C, but a peak value of 70 mg/kg DM at 500°C. The values decreased strongly again at higher temperatures, with a minimum concentration of 0.18 mg/kg DM at 700°C. In contrast, Dai et al. (2014) found increased PAH concentrations at increased temperature, with a maximum at 950°C, but they identified other factors such as gas flow, sample mass and residence time as important factors as well. This illustrates that, in addition to differences caused by varying feedstock, the PAH concentration largely depends on pyrolysis conditions. In general, pyrolysis resulted in a decreased proportion of 5- and 6-ring compounds in the PAH mixture, whereas the proportion of the less toxic naphthalene increased (Paz-Ferreiro et al., 2018).

At higher temperatures (> 1000°C) and under optimized thermal processes, potentially toxic organic compounds can be destroyed efficiently. For example, pyrolysis of sewage sludge at 1400°C in a melting furnace resulted in destruction of 99.9% of dioxins (Lindberg et al., 2015). Generally, only low concentrations of dioxins or furans can occur under oxygen-depleted conditions as both of these groups of compounds contain oxygen. Consequently, concentrations of dioxins/furans in carbonaceous materials were often below the limit of detection (Buss, 2016).

Only limited data are available about volatile organic compounds (VOC) that can also be formed during pyrolysis; they have high mobility and therefore strong biological activity (Buss, 2016).

In Table 8 organic contaminations in recycled products are compiled from literature. Empty cells denote missing values and should not be mistaken for an absence of contamination. The data in Table 8 are based on a low number of studies, but indicate a general decrease in the organic contamination in some of the materials shown when compared to the data in Table 5. The data availability is too poor to draw general conclusions about organic contaminations for ash-based and pyrolysed materials; however, products such as struvite appear to be a promising alternative fertilizer with low concentration of organic pollutants.

Table 8 Organic contaminants in treated-waste fertilizers

Organic contaminant	Struvite	Materials based on sewage sludge ash	Carbonaceous materials based on sewage sludge
Σ of 16 PAHs (mg/kg DM); *Benzo(a)pyrene: 1 mg/ kg DM	<LOD-0.6 (n = 8)[d] <0.09-2.6 (n = 3)[c] 5.7-15 (n = 3)[e]	0.97-17.5 (n = 10 raw ash)[e] 2.1-15 (n = 6 fertilizers from treated ash)[e]	0.96 ± 0.062[g] 1.43 ± 0.04[i] 0.18-70[b] 21-37 (n = 5)[f]
PCBs (ng/kg DM) *per compound: 0.1 mg/ kg DM	358 (n = 1)[a]	855 (n = 1)[a]	
Dioxin-like (dl) PCBs (ng WHO-TEQ/kg DM)	0.7-0.9 (n = 3)[c]		
PCDD/F (ng/kg DM)	15 (n = 1)[a] 0.82 - >223 (n = 3)[c]	<LOD (n = 1)[a]	
PCDD/F (ng WHO-TEQ/ kg DM)	1.5-1.9 (n = 8)[d] 1.5-2.4 (n = 3)[c]		
ΣPCDD/F + dl-PCB (ng WHO-TEQ/kg DM) **ΣPCDD/F + dl-PCB 30 (8) ng WHO-TEQ/kg DM	0.28-3.41 (n = 3)[e]	0.01-8.09 (n = 10 raw ash)[e] 0.07-1.82 (n = 6 fertilizers from treated ash)[e]	often <LOD[f]
AOX (mg/kg DM) *Σ of all AOX 400 mg/kg DM	<71.3-<85.1 (n = 3)[c]		
Other compounds (ng/g DM)	Carbamazepin 0.4-113[h] Benzotriazol 6.9-108[h] Diclofenac 1.1-8.5[h]		VOCs – volatile organic compounds can be formed[f]

*Limit value according to the German Sewage Sludge Ordinance (AbfKlärV).
**Limit value according to the German Fertilizer Ordinance (DüMV) (8 ng WHO-TEQ/kg DM for feed production fields).
LOD – lower limit of detection.
References: [a]Analyses by the chapter authors, [b]Luo et al. (2014), [c]Gerhardt et al. (2015), [d]Kraus and Seis (2015), [e]Zeggel et al. (2015), [f]Buss (2016), [g]De la Rosa et al. (2016), [h]Rastetter et al. (2017), [i]Fristak et al. (2018).

5 Future trends and conclusion

Recycling of organic waste materials and closing nutrient cycles are indispensable in preventing uncontrolled nutrient losses to natural systems and ensuring a sufficient nutrient supply to plants in the future especially with respect to P nutrition. It is generally accepted that sewage sludge or composts are valuable sources of (i) nutrients and (ii) SOM with the potential to improve the soil water infiltration and holding capacity as well as plant yields, especially on poor sandy soils low in organic matter. However, these valuable organic nutrient sources can have unacceptably high concentrations of organic and inorganic pollutants and other harmful substances or organisms, as discussed above.

Furthermore, sewage sludge and other human wastes typically accumulate in areas of high population density and their nutrients must be distributed to regions where they are required in agriculture. However, transport of materials such as sewage sludge is often not economic because of their high water content (Bloem et al., 2017), and they are mostly distributed where produced, resulting in environmental problems such as eutrophication. Transport may become economic after drying and/or incineration of the sewage sludge as the volume is reduced by as much as 70% (Fytili and Zabaniotou, 2008).

There is no single optimum solution for all agricultural regions of the world because the challenges different regions face are diverse.

Waste materials can be valuable fertilizer products if harmful substances are removed or their concentration decreased, and providing their nutrients in plant-available forms. Different strategies to reach this goal were comprehensively discussed above, especially for residues from municipal waste water treatment. Due to its limited recovery rates, precipitation of N and P on site offers only a partial solution. Assuming an optimized thermal process that would produce ashes almost free of carbon and consequently organic pollutants, P remaining in the highly contaminated sludge should be recovered by mono-incineration combined with a post-treatment to decrease heavy metal concentrations and increase plant availability of P.

One future possibility to further improve the exploitation of nutrients from waste water could be alternative waste water systems, such as source separation of urine, because more than 90% of N and 50% of P in the domestic waste water originates from urine (Larsen et al., 2001; De Boer et al., 2018). Also, to minimize contamination, new approaches to waste water treatment would be advantageous, including separation of industrial and domestic waste water, as well as separation and recycling of grey water (water from showers and washing basins), for example flushing toilets. Such measures help save water, an important aspect regarding climate change, and would decrease the total amount of waste water. Contamination with pharmaceuticals was reduced to <1% when struvite was precipitated from urine directly in a 2-stage process whereby spontaneous precipitation after addition of urease was followed by controlled crystallization of struvite by Mg addition after a 5-day storage (Kemacheevakul et al., 2014). Tirado and Allsopp (2012) described the implementation of ecological sanitation systems, which also rely on separating waste streams, as the best solution to recovering nutrients from human excreta, thereby closing nutrient cycles and at the same time saving water and energy. Still, a half of the world population has no access to sanitation facilities, which can be seen as a chance to implement more sustainable systems in the future than the 'flush and forget' systems generally built in industrialized countries and wasting water and diluting nutrients. Transport of urine is more economic than transport of sewage sludge because of the higher nutrient concentration.

For Sweden, it was estimated that transport of urine up to 100 km remains more energy efficient than production and use of the equivalent amount of mineral P fertilizers (Cordell, 2010).

A target of future research and modern sanitation systems should be to facilitate safe reuse of organic matter because it is needed in many regions of the world to improve soil functions and is also beneficial with respect to C sequestration. A possible approach to retain organic matter would be an advanced separation of clean and contaminated sources. Source separation and separation of clean and contaminated streams would reduce the total amounts of contaminated waste materials. This could help develop adequate treatment options, such as longer composting durations for the degradation of harmful organic compounds.

The treatment options discussed in this chapter are promising and show that it is already possible to produce fertilizers from treated wastes that can at least partly replace mineral P fertilizers. However, the following major targets for the future with respect to optimizing the use of treated wastes in agriculture remain:

- recovery of waste-bound nutrients in plant-available forms,
- minimizing contamination of wastes, including inorganic and organic contaminants as well as other harmful components such as prions, pathogens and resistant bacteria, and
- optimum distribution of nutrient sources to farmland to provide nutrients where needed and prevent unwanted losses to the environment.

These major targets could be further supported by adapting existing legal regulations. One important step to decreasing contamination would be to replace concentration-based limit values by pollutant-to-nutrient ratios because this ratio determines the flow of pollutants to the soil. New emerging pollutants such as microplastics are an additional challenge when using waste materials. Therefore, eco-toxicological methods should be further developed and included in future regulations as they can be early-warning systems and can help evaluate new fertilizer products made from waste materials. It is important to have a set of methods that can be used for different materials (liquid as well as solid fertilizer sources) and for mirroring the ecosystems with their complex interactions among soil, water, plants, microorganisms and humans.

6 Where to look for further information

Optimizing the use of treated wastes in crop nutrition is a topic which is relevant across the borders. Therefore, several platforms exist which cover a broad range of the topics that are important with respect to a sustainable nutrient utilization in a united world. They deliver valuable and up-to-date information on nutrient utilization from several treated wastes, including sewage sludge as well as manure, domestic biowastes and industrial organic wastes. In addition to the references already pointed out in this chapter, the authors would like to refer the reader to the following links to look for further information from a European perspective:

Biorefine Cluster Europe, interconnecting projects and people within the domain of bio-based resource recovery (www.biorefine.eu).

SuMaNu HELCOM, a project platform gathering and synthesizing the best practices and recommendations on sustainable manure and nutrient management (www.helcom.fi/helcom-at-work/projects/sumanu).

European Sustainable Phosphorus Platform (ESPP), facilitating knowledge exchange and discussion between market, stakeholders and regulators, and circulating information on sustainable P management in agriculture, food, industry, water and environment (https://phosphorusplatform.eu).

7 References

Abdel-Shafy, H. I. and Mansour, M. S. M. 2016. A review on polycyclic aromatic hydrocarbons: source, environmental impact, effect on human health and remediation. *Egyptian Journal of Petroleum* 25(1), 107–23. doi:10.1016/j.ejpe.2015.03.011.

Abwasserverband Braunschweig. 2014. Kombinationsverfahren zur Rückgewinnung von Phosphor und Stickstoff bei gleichzeitiger Energieoptimierung der Kläranlage. Available at: https://www.abwasserverband-bs.de/wp-content/uploads/2014/06/Projektbeschreibung-N%C3%A4hrstoffr%C3%BCckgewinnung-AVBS.pdf (accessed on 9 January 2019).

Adam, C. 2018. Verfahren zur Phosphorrückgewinnung aus Abwasser und Klärschlamm. In: Holm, O., Thomé-Kozmiensky, E., Quicker, P. and Kopp-Assenmacher, S. (Eds), *Verwertung von Klärschlamm*. Tagungsband zur Berliner Klärschlammkonferenz 5./6. November 2018, pp. 165–84. ISBN 978-3-944310-43-5.

Adam, C., Peplinski, B., Michaelis, M., Kley, G. and Simon, F. G. 2009. Thermochemical treatment of sewage sludge ashes for phosphorus recovery. *Waste Management* 29(3), 1122–8. doi:10.1016/j.wasman.2008.09.011.

Agegnehu, G., Srivastava, A. K. and Bird, M. I. 2017. The role of biochar and biochar-compost in improving soil quality and crop performance: a review. *Applied Soil Ecology* 119, 156–70. doi:10.1016/j.apsoil.2017.06.008.

Ahmed, M. B., Zhou, J. L., Ngo, H. H., Guo, W., Thomidis, N. S. and Xu, J. 2017. Progress in the biological and chemical treatment technologies for emerging contaminant removal from wastewater: a critical review. *Journal of Hazardous Materials* 323(A), 274-98. doi:10.1016/j.jhazmat.2016.04.045.

Ahmed, N., Shim, S., Won, S. and Ra, C. 2018. Struvite recovered from various types of wastewaters: characteristics, soil leaching behavior, and plant growth. *Land Degradation and Development* 29(9), 2864–79. doi:10.1002/ldr.3010.

Aller, M. F. 2016. Biochar properties: transport, fate, and impact. *Critical Reviews in Environmental Science and Technology* 46(14–15), 1183–296. doi:10.1080/106433 89.2016.1212368.

Alvarino, T., Suarez, S., Lema, J. and Omil, F. 2018. Understanding the sorption and biotransformation of organic micropollutants in innovative biological wastewater treatment technologies. *The Science of the Total Environment* 615, 297–306. doi:10.1016/j.scitotenv.2017.09.278.

Angst, T. E. and Sohi, S. P. 2013. Establishing release dynamics for plant nutrients from biochar. *GCB Bioenergy* 5(2), 221–6. doi:10.1111/gcbb.12023.

Antakyali, D., Meyer, C., Preyl, V., Maier, W. and Steinmetz, H. 2013. Large-scale application of nutrient recovery from digested sludge as struvite. *Water Practice and Technology* 8(2), 256–62. doi:10.2166/wpt.2013.027.

Antille, D. L., Sakrabani, R. and Godwin, R. J. 2013. Field-scale evaluation of biosolids-derived organomineral fertilizers applied to ryegrass (*Lolium perenne* L.) in England. *Applied and Environmental Soil Science* 2013, Article 960629.

Antille, D. L., Sakrabani, R. and Godwin, R. J. 2014. Effects of biosolids-derived organomineral fertilizers, urea and biosolids granules on crop and soil established with ryegrass (*Lolium perenne* L.). *Communications in Soil Science and Plant Analysis* 45(12), 1605–21. doi:10.1080/00103624.2013.875205.

Antille, D. L., Godwin, R. J., Sakrabani, R., Seneweera, S., Tyrrel, S. F. and Johnston, A. E. 2017. Field-scale evaluation of biosolids-derived organomineral fertilizers applied to winter wheat in England. *Agronomy Journal* 109(2), 654–74. doi:10.2134/agronj2016.09.0495.

Appel, T. and Friedrich, K. 2017. *Phosphor-Recycling mit Karbonisaten aus Klärschlamm*. Schriftenreihe des Hermann-Hoepke-Instituts, Band 1. TH Bingen, Bingen, Germany, 242p. ISBN 978-3-9810496-2-6.

Appel, T., Friedrich, K., Susset, D. and Pint, F. 2016. Soda-Additiv beim Karbonisieren von Klärschlamm steigert die Phosphor-Düngewirkung im Gefäßversuch mit Mais. VDLUFA Schriftenreihe 73, Kongressband 2016, pp. 362–8.

Archontoulis, S. V., Huber, I., Miguez, F. E., Thorburn, P. J., Rogavska, N. and Laird, D. A. 2016. A model for mechanistic and system assessments of biochar effects on soils and crops and trade-offs. *GCB Bioenergy* 8(6), 1028–45. doi:10.1111/gcbb.12314.

Bergé, A., Cladière, M., Gasperi, J., Coursimault, A., Tassin, B. and Moilleron, R. 2012. Meta-analysis of environmental contamination by alkylphenols. *Environmental Science and Pollution Research International* 19(9), 3798–819. doi:10.1007/s11356-012-1094-7.

Biedermann, L. A. and Harpole, W. S. 2013. Biochar and its effects on plant productivity and nutrient cycling: a meta-analysis. *GCB Bioenergy* 5(2), 202–14. doi:10.1111/gcbb.12037.

Bloem, E., Albihn, A., Elving, J., Hermann, L., Lehmann, L., Sarvi, M., Schaaf, T., Schick, J., Turtola, E. and Ylivainio, K. 2017. Contamination of organic nutrient sources with potentially toxic elements, antibiotics and pathogen microorganisms in relation to P fertilizer potential and treatment options for the production of sustainable fertilizers: a review. *The Science of the Total Environment* 607–608, 225–42. doi:10.1016/j.scitotenv.2017.06.274.

Boehler, M. A., Heisele, A., Seyfried, A., Grömping, M. and Siegrist, H. 2015. $(NH_4)_2SO_4$ recovery from liquid side streams. *Environmental Science and Pollution Research International* 22(10), 7295–305. doi:10.1007/s11356-014-3392-8.

Bogner, R. and Ortwein, B. 2018. Phosphorrückgewinnung – Geeignete Verfahren hinsichtlich der neuen Anforderungen. In: Holm, O., Thomé-Kozmiensky, E., Quicker, P. and Kopp-Assenmacher, S. (Eds), *Verwertung von Klärschlamm*. Tagungsband zur Berliner Klärschlammkonferenz 5/6. November 2018, pp. 419-26. ISBN 978-3-944310-43-5.

Börjesson, G. and Kätterer, T. 2018. Soil fertility effects of repeated application of sewage sludge in two 30-year-old field experiments. *Nutrient Cycling in Agroecosystems* 112(3), 369-85. doi:10.1007/s10705-018-9952-4.

Brown, K., Harrison, J. and Bowers, K. 2018. Struvite precipitation from anaerobically digested dairy manure. *Water, Air, and Soil Pollution* 229(7), 217. doi:10.1007/s11270-018-3855-5.

Bruun, S., Harmer, S. L., Bekiaris, G., Christel, W., Zuin, L., Hu, Y., Jensen, L. S. and Lombi, E. 2017. The effect of different pyrolysis temperatures on the speciation and availability in soil of P in biochar produced from the solid fraction of manure. *Chemosphere* 169, 377-86. doi:10.1016/j.chemosphere.2016.11.058.

Buss, W. 2016. Contaminant issues in production and application of biochar. Dissertation. The University of Edinburgh. Available at: https://www.era.lib.ed.ac.uk/handle/1842/25526 (accessed on 24 January 2019).

Butnan, S., Deenik, J. L., Toomsan, B., Antal, M. J. and Vityakon, P. 2015. Biochar characteristics and application rates affecting corn growth and properties of soils contrasting in texture and mineralogy. *Geoderma* 237-238, 105-16. doi:10.1016/j.geoderma.2014.08.010.

Cabeza, R., Steingrobe, B., Römer, W. and Claassen, N. 2011. Effectiveness of recycled P products as P fertilizers, as evaluated in pot experiments. *Nutrient Cycling in Agroecosystems* 91(2), 173-84. doi:10.1007/s10705-011-9454-0.

Cascarosa, E., Gea, G. and Arauzo, J. 2012. Thermochemical processing of meat and bone meal: a review. *Renewable and Sustainable Energy Reviews* 16(1), 942-57. doi:10.1016/j.rser.2011.09.015.

CCME. 2010. A review of the current Canadian Legislative Framework for Wastewater Biosolids. Canadian Council of Ministers of the Environment. PN 1446. Available at: https://www.ccme.ca/files/Resources/waste/biosolids/pn_1446_biosolids_leg_r eview:eng.pdf (accessed on 24 January 2019).

Chang, S. C. and Jackson, M. L. 1957. Fractionation of soil phosphorus. *Soil Science* 84(2), 133-44. doi:10.1097/00010694-195708000-00005.

Chen, Q., An, X., Li, H., Su, J., Ma, Y. and Zhu, Y. G. 2016. Long-term field application of sewage sludge increases the abundance of antibiotic resistance genes in soil. *Environment International* 92-93, 1-10. doi:10.1016/j.envint.2016.03.026.

Cincinelli, A., Martellini, T., Misuri, L., Lanciotti, E., Sweetman, A., Laschi, S. and Palchetti, I. 2012. PBDEs in Italian sewage sludge and environmental risk of using sewage sludge for land application. *Environmental Pollution* 161, 229-34. doi:10.1016/j.envpol.2011.11.001.

Clarke, B. O. and Smith, S. R. 2011. Review of 'emerging' organic contaminants in biosolids and assessment of international research priorities for the agricultural use of biosolids. *Environment International* 37(1), 226-47. doi:10.1016/j.envint.2010.06.004.

Cordell, D. 2010. The story of phosphorus: sustainability implications of global phosphorus scarcity for food security. PhD Thesis; Linköping Studies in Arts and Science No. 509. Department of Water and Environmental Studies, Linköping University.

Cordell, D. and White, S. 2011. Peak phosphorus: clarifying the key issues of a vigorous debate about long-term phosphorus security. *Sustainability* 3(10), 2027-49. doi:10.3390/su3102027.

Crane-Droesch, A., Abiven, S., Jeffery, S. and Torn, M. S. 2013. Heterogeneous global crop yield response to biochar: a meta-regression analysis. *Environmental Research Letters* 8(4). doi:10.1088/1748-9326/8/4/044049.

Crombie, K., Mašek, O., Cross, A. and Sohi, S. 2015. Biochar – synergies and trade-offs between soil enhancing properties and C sequestration potential. *GCB Bioenergy* 7(5), 1161–75. doi:10.1111/gcbb.12213.

Dai, Q., Jiang, X., Jiang, Y., Jin, Y., Wang, F., Chi, Y. and Yan, J. 2014. Formation of PAHs during the pyrolysis of dry sewage sludge. *Fuel* 130, 92–9. doi:10.1016/j.fuel.2014.04.017.

Dai, L., Li, H., Tan, F., Zhu, N., He, M. and Hu, G. 2016. Biochar: a potential route for recycling of phosphorus in agricultural residues. *GCB Bioenergy* 8(5), 852–8. doi:10.1111/gcbb.12365.

Day, D., Evens, R. J., Lee, J. W. and Reicosky, D. 2005. Economical CO_2, SO_x, and NO_x capture from fossil-fuel utilization with combined renewable hydrogen production and large-scale carbon sequestration. *Energy* 30(14), 2558–79. doi:10.1016/j.energy.2004.07.016.

De Boer, M. A., Hammerton, M. and Slootweg, J. C. 2018. Uptake of pharmaceuticals by sorbent-amended struvite fertilisers recovered from human urine and their bioaccumulation in tomato fruit. *Water Research* 133, 19–26. doi:10.1016/j.watres.2018.01.017.

Degryse, F., Baird, R., Da Silva, R. C. and McLaughlin, M. J. 2017. Dissolution rate and agronomic effectiveness of struvite fertilizers – effect of soil pH, granulation and base excess. *Plant and Soil* 410(1–2), 139–52. doi:10.1007/s11104-016-2990-2.

De Kok, L. and Schnug, E. (Eds). 2008. *Loads and Fate of Fertilizer Derived Uranium*. Backhuys Publishers, Leiden, The Netherlands, 229pp. ISBN 978-3-8236-1546-0.

De la Rosa, J. M., Paneque, M., Hilber, I., Blum, F., Knicker, H. E. and Bucheli, T. D. 2016. Assessment of polycyclic aromatic hydrocarbons in biochar and biochar-amended agricultural soil from Southern Spain. *Journal of Soils and Sediments* 16(2), 557–65. doi:10.1007/s11368-015-1250-z.

Delin, S. 2016. Fertilizer value of phosphorus in different residues. *Soil Use and Management* 32, 17–26.

Demirbaş, A. 2001. Carbonization ranking of selected biomass for charcoal, liquid and gaseous products. *Energy Conversion and Management* 42(10), 1229–38. doi:10.1016/S0196-8904(00)00110-2.

Duboc, O., Santner, J., Golestani Fard, A., Zehetner, F., Tacconi, J. and Wenzel, W. W. 2017. Predicting phosphorus availability from chemically diverse conventional and recycling fertilizers. *The Science of the Total Environment* 599–600, 1160–70. doi:10.1016/j.scitotenv.2017.05.054.

Egle, L., Rechberger, H. and Zessner, M. 2015. Overview and description of technologies for recovering phosphorus from municipal wastewater. *Resources, Conservation and Recycling, Part B* 105, 325–46.

Ehbrecht, A., Ritter, H. J., Schmidt, S. O., Schönauer, S., Schuhmann, R. and Weber, N. 2016. Entwicklung eines kombinierten Kristallisationsverfahrens zur Gewinnung von Phosphatdünger aus dem Abwasserreinigungsprozess mit vollständiger Verwertung der Restphasen in der Zementindustrie. Abschlussbericht zum Forschungsvorhaben AiF-Nr. 17899 N. Forschungsbericht Nr. 2/2016. Available at: http://www.fg-kalk-m oertel.de/files/02_2016_Forschungsbericht_Phosphatrecycling.pdf (accessed on 24 January 2019).

Epstein, E. 2003. *Land Application of Sewage Sludge and Biosolids*. Lewis Publishers, Boca Raton, FL, 201pp. ISBN 1-56670-624-6.

European Commission. 2014a. Towards a circular economy: a zero waste programme for Europe. COM 2014 398 final. Available at: http://ec.europa.eu/transparency/re gdoc/rep/1/2014/EN/1-2014-398-EN-F1-1.Pdf (accessed on 24 January 2019).

European Commission. 2014b. On the review of the list of critical raw materials for the EU and the implementation of the Raw Materials Initiative. COM 2014 297 final. Available at: https://www.kowi.de/Portaldata/2/Resources/fp/2014-COM-Critical-R aw-Materials-List.pdf (accessed on 24 January 2019).

European Commission. 2015. Closing the loop - an EU action plan for the Circular Economy. COM 2015 614 final. Available at: https://eur-lex.europa.eu/resource. html?uri=cellar:8a8ef5e8-99a0-11e5-b3b7-01aa75ed71a1.0012.02/DOC_1&for mat=PDF (accessed on 24 January 2019).

European Commission. 2016. Sewage sludge. Last updated 08/06/2016. Available at: http://ec.europa.eu/environment/waste/sludge/index.htm (accessed on 24 January 2019).

European Commission. 2017. 2017 list of Critical Raw Materials for the EU. COM 2017 490 final. Available at: https://eur-lex.europa.eu/legal-content/EN/TXT/PDF/?uri =CELEX:52017DC0490&from=EN (accessed on 24 January 2019).

EUROSTAT. 2018. Sewage sludge production and disposal. Last updated 9 April 2018. Available at: https://ec.europa.eu/eurostat/web/products-datasets/product?code= env_ww:spd (accessed on 15 January 2019).

Fijalkowski, K., Rorat, A., Grobelak, A. and Kacprzak, M. J. 2017. The presence of contaminants in sewage sludge - the current situation. *Journal of Environmental Management* 203(3), 1126-36. doi:10.1016/j.jenvman.2017.05.068.

Fornefeld, E. and Smalla, K. 2018. Hygienische Risiken bei der landwirtschaftlichen Verwertung von Klärschlamm durch die Aufnahme von Infektionserregern in Kulturpflanzen. (Abschlussbericht) Umwelt & Gesundheit 03/2018, Umweltbundesamt. Available at: https://www.umweltbundesamt.de/sites/default/ files/medien/1410/publikationen/2018-06-21_umwelt-und-gesundheit_03-2018_ hygienische-risiken-klaerschlamm.pdf (accessed on 24 January 2019).

Fristak, V., Pipiska, M. and Soja, G. 2018. Pyrolysis treatment of sewage sludge: a promising way to produce phosphorus fertilizer. *Journal of Cleaner Production* 172, 1772-8. doi:10.1016/j.jclepro.2017.12.015.

Frossard, E., Tekely, P. and Grimal, J. Y. 1994. Characterization of phosphate species in urban sewage sludges by high-resolution solid-state ^{31}P NMR. *European Journal of Soil Science* 45(4), 403-8. doi:10.1111/j.1365-2389.1994.tb00525.x.

Frossard, E., Sinaj, S. and Dufour, P. 1996. Phosphorus in urban sewage sludges as assessed by isotopic exchange. *Soil Science Society of America Journal* 60(1), 179-82. doi:10.2136/sssaj1996.03615995006000010029x.

Frossard, E., Bauer, J. P. and Lothe, F. 1997. Evidence of vivianite in $FeSO_4$-flocculated sludges. *Water Research* 31(10), 2449-54. doi:10.1016/S0043-1354(97)00101-2.

Frost, P., Camenzind, R., Mägert, A., Bonjour, R. and Karlaganis, G. 1993. Organic micropollutants in Swiss sewage sludge. *Journal of Chromatography* 643(1-2), 379-88. doi:10.1016/0021-9673(93)80574-R.

Fytili, D. and Zabaniotou, A. 2008. Utilization of sewage sludge in EU application of old and new methods-a review. *Renewable and Sustainable Energy Reviews* 12(1), 116-40. doi:10.1016/j.rser.2006.05.014.

Gaskin, J., Steiner, C., Harris, K., Das, K. and Bibens, B. 2008. Effect of low-temperature pyrolysis conditions on biochar for agricultural use. *Transactions of the ASABE* 51(6), 2061–9. doi:10.13031/2013.25409.

Gaunt, J. L. and Lehmann, J. 2008. Energy balance and emissions associated with biochar sequestration and pyrolysis bioenergy production. *Environmental Science and Technology* 42(11), 4152–8. doi:10.1021/es071361i.

Gerhardt, A., Kabbe, C., Rastetter, N., Stemann, J. and Wilken, V. 2015. Quantification of nutritional value and toxic effects of each P recovery product, Deliverable D 8.1 of the EU P-REX Project. Available at: https://zenodo.org/record/242550#.XEqyLT NCe70 (accessed on 24 January 2019).

Goldbach, H. E., Clemens, J. and Scherer, H. W. 2007. Nährstoffgehalte und –wirkungen verschiedener Klärschlämme – Auswirkungen langjähriger Klärschlammdüngung auf die Bodenfruchtbarkeit. In: Kuratorium für Technik und Bauwesen in der Landwirtschaft (KTBL) (Ed.), *Perspektiven der Klärschlammverwertung. Ziele und Inhalte einer Novelle der Klärschlammverordnung.* KTBL-Schrift 453. KTBL Darmstadt, pp. 171–81. ISBN 978-3-939371-23-6.

Gonzaga, M. I. S., Mackowiak, C. L., Comerford, N. B., Moline, E. FdV., Shirley, J. P. and Guimaraes, D. V. 2017. Pyrolysis methods impact biosolids-derived biochar composition, maize growth and nutrition. *Soil and Tillage Research* 165, 59–65. doi:10.1016/j.still.2016.07.009.

Grandclément, C., Seyssiecq, I., Piram, A., Wong-Wah-Chung, P., Vanot, G., Tiliacos, N., Roche, N. and Doumenq, P. 2017. From the conventional biological wastewater treatment to hybrid processes, the evaluation of organic micropollutant removal: a review. *Water Research* 111, 297–317. doi:10.1016/j.watres.2017.01.005.

Gul, S. and Whalen, J. K. 2016. Biochemical cycling of nitrogen and phosphorus in biochar-amended soils. *Soil Biology and Biochemistry* 103, 1–15. doi:10.1016/j.soilbio.2016.08.001.

Hagspiel, B. 2018. RN-Mephrec - Klärschlammverwertung Region Nürnberg - Klärschlamm zu Energie, Dünger und Eisen mit metallurgischem Phosphorrecycling in einem Verfahrensschritt. BMBF (02WER1313A). Available at: https://bmbf.na wam-erwas.de/de/project/krn-mephrec (accessed on 25 January 2019).

Hamdi, H., Hechmi, S., Khelil, M. N., Zoghlami, I. R., Benzarti, S., Mokni-Tlili, S., Hassen, A. and Jedidi, N. 2019. Repetitive land application of urban sewage sludge: effect of amendment rates and soil texture on fertility and degradation parameters. *Catena* 172, 11–20. doi:10.1016/j.catena.2018.08.015.

Harrison, E. Z., Oakes, S. R., Hysell, M. and Hay, A. 2006. Organic chemicals in sewage sludges. *The Science of the Total Environment* 367(2-3), 481–97. doi:10.1016/j.scitotenv.2006.04.002.

He, P., Liu, Y., Shao, L., Zhang, H. and Lü, F. 2018. Particle size dependence of the physicochemical properties of biochar. *Chemosphere* 212, 385–92. doi:10.1016/j.chemosphere.2018.08.106.

Hedley, M. J., Stewart, J. W. B. and Chauhan, B. S. 1982. Changes in inorganic and organic soil phosphorus fractions induced by cultivation practices and by laboratory incubations. *Soil Science Society of America Journal* 46(5), 970–6. doi:10.2136/sssaj1982.03615995004600050017x.

Hemmat, A., Aghilinategh, N., Rezainejad, Y. and Sadeghi, M. 2010. Long-term impacts of municipal solid waste compost, sewage sludge and farmyard manure application on organic carbon, bulk density and consistency limits of a calcareous soil in central Iran. *Soil and Tillage Research* 108(1–2), 43–50. doi:10.1016/j. still.2010.03.007.

Herzel, H., Krüger, O., Hermann, L. and Adam, C. 2016. Sewage sludge ash – a promising secondary phosphorus source for fertilizer production. *The Science of the Total Environment* 542(B), 1136–43. doi:10.1016/j.scitotenv.2015.08.059.

Hoffman, T. C., Zitomer, D. H. and McNamara, P. J. 2016. Pyrolysis of wastewater biosolids significantly reduces estrogenicity. *Journal of Hazardous Materials* 317, 579–84. doi:10.1016/j.jhazmat.2016.05.088.

Hossain, M. K., Strezov, V., Chan, K. Y., Ziolkowski, A. and Nelson, P. F. 2011. Influence of pyrolysis temperature on production and nutrient properties of wastewater sludge biochar. *Journal of Environmental Management* 92(1), 223–8. doi:10.1016/j. jenvman.2010.09.008.

Huang, X. L. and Shenker, M. 2004. Water-soluble and solid-state speciation in stabilized sewage sludge. *Journal of Environmental Quality* 33(5), 1895–903.

Huang, R. and Tang, Y. 2015. Speciation dynamics of phosphorus during (hydro)thermal treatments of sewage sludge. *Environmental Science and Technology* 49(24), 14466–74. doi:10.1021/acs.est.5b04140.

Huang, R. and Tang, Y. 2016. Evolution of phosphorus complexation and mineralogy during (hydro)thermal treatments of activated and anaerobically digested sludge: insights from sequential extraction and P K-edge XANES. *Water Research* 100, 439–47. doi:10.1016/j.watres.2016.05.029.

Huang, X. L., Chen, Y. and Shenker, M. 2012. Dynamics of phosphorus bioavailability in soil amended with stabilized sewage sludge materials. *Geoderma* 170, 144–53. doi:10.1016/j.geoderma.2011.11.025.

Huerta-Fontela, M., Galceran, M. T. and Ventura, F. 2010. Fast liquid chromatography-quadrupole-linear ion trap mass spectrometry for the analysis of pharmaceuticals and hormones in water resources. *Journal of Chromatography A* 1217(25), 4212–22. doi:10.1016/j.chroma.2009.11.007.

Hussain, M., Farooq, M., Nawaz, A., Al-Sadi, A. M., Solaiman, Z. M., Alghamdi, S. S., Ammara, U., Ok, Y. S. and Siddique, K. H. M. 2017. Biochar for crop production: potential benefits and risks. *Journal of Soils and Sediments* 17(3), 685–716. doi:10.1007/s11368-016-1360-2.

Ibarrola, R., Shackley, S. and Hammond, J. 2012. Pyrolysis biochar systems for recovering biodegradable materials: a life cycle carbon assessment. *Waste Management* 32(5), 859–68. doi:10.1016/j.wasman.2011.10.005.

Jaafar, N. M., Clode, P. L. and Abbott, L. K. 2015. Soil microbial response to biochars varying in particle size, surface and pore properties. *Pedosphere* 25(5), 770–80. doi:10.1016/S1002-0160(15)30058-8.

Jeffery, S., Verheijen, F. G. A., van der Velde, M. and Bastos, A. C. 2011. A quantitative review of the effects of biochar application to soils on crop productivity using meta-analysis. *Agriculture, Ecosystems and Environment* 144(1), 175–87. doi:10.1016/j. agee.2011.08.015.

Jeffery, S., Bezemer, T. M., Cornelissen, G., Kuyper, T. W., Lehmann, J., Mommer, L., Sohi, S. P., van de Voorde, T. F. J., Wardle, D. A. and van Groenigen, J. W. 2015. The way forward in biochar research: targeting trade-offs between the potential wins. *GCB Bioenergy* 7(1), 1–13. doi:10.1111/gcbb.12132.

Jensen, A., Dam-Johansen, K., Wójtowicz, M. A. and Serio, M. A. 1998. TG-FTIR study of the influence of potassium chloride on wheat straw pyrolysis. *Energy and Fuels* 12(5), 929–38. doi:10.1021/ef980008i.

Jones, V., Gardner, M. and Ellor, B. 2014. Concentrations of trace substances in sewage sludge from 28 wastewater treatment works in the UK. *Chemosphere* 111, 478–84. doi:10.1016/j.chemosphere.2014.04.025.

JRC. 2001. Organic contaminants in sewage sludge for agricultural use. Available at: http://ec.europa.eu/environment/archives/waste/sludge/pdf/organics_in_sludge.pdf (accessed on 24 January 2019).

Kahiluoto, H., Kuisma, M., Ketoja, E., Salo, T. and Heikkinen, J. 2015. Phosphorus in manure and sewage sludge more recyclable than in soluble inorganic fertilizer. *Environmental Science and Technology* 49(4), 2115–22. doi:10.1021/es503387y.

Kapanen, A., Vikman, M., Rajasärkkä, J., Virta, M. and Itävaara, M. 2013. Biotests for environmental quality assessment of composted sewage sludge. *Waste Management* 33(6), 1451–60. doi:10.1016/j.wasman.2013.02.022.

Karbalaei, S., Hanachi, P., Walker, T. R. and Cole, M. 2018. Occurrence, sources, human health impacts and mitigation of microplastic pollution. *Environmental Science and Pollution Research International* 25(36), 36046–63. doi:10.1007/s11356-018-3508-7.

Kemacheevakul, P., Chuangchote, S., Otani, S., Matsuda, T. and Shimizu, Y. 2014. Phosphorus recovery: minimization of amount of pharmaceuticals and improvement of purity in struvite recovered from hydrolysed urine. *Environmental Technology* 35(21–24), 3011–9. doi:10.1080/09593330.2014.929179.

Kern, D., Friedrich, K. and Appel, T. 2016. Phosphor-Düngewirkung und Schwermetallverfügbarkeit von karbonisiertem Klärschlamm in Abhängigkeit von der Dauer der Karbonisierung und der Zugabe von Natriumsulfat. VDLUFA-Schriftenreihe 73, Kongressband 2016, pp. 355–61.

Kirchmann, H., Börjesson, G., Kätterer, T. and Cohen, Y. 2017. From agricultural use of sewage sludge to nutrient extraction: a soil science outlook. *Ambio* 46(2), 143–54. doi:10.1007/s13280-016-0816-3.

Koch, M., Adam, C. and Krüger, O. 2018. Untersuchung der Anwendbarkeit der im Rahmen des CEN-Projekts HORIZONTAL entwickelten Analyseverfahren auf Düngemittel und Klärschlamm/-aschen. UBA-Texte 3/2018. Available at: https://www.umweltbundesamt.de/sites/default/files/medien/1410/publikationen/2018-01-10_texte_03-2018_anwendbarkeit-horizontal.pdf (accessed on 24 January 2019).

Kominko, H., Gorazda, K. and Wzorek, Z. 2017. The possibility of organo-mineral fertilizer production from sewage sludge. *Waste and Biomass Valorization* 8(5), 1781–91. doi:10.1007/s12649-016-9805-9.

Kratz, S., Haneklaus, S. and Schnug, E. 2010. Chemical solubility and agricultural performance of P-containing recycling fertilizers. *Landbauforschung vTI Agriculture and Forestry Research* 60, 227–40.

Kratz, S., Schick, J. and Øgaard, A. F. 2016. P solubility of inorganic and organic P sources. In: Schnug, E. and de Kok, L. (Eds), *Phosphorus: 100% Zero*. Springer, Berlin, Heidelberg, pp. 127–54. ISBN 978-94-017-7611-0.

Kratz, S., Bloem, E., Papendorf, J., Schick, J., Schnug, E. and Harborth, P. 2017. Agronomic efficiency and heavy metal contamination of phosphorus (P) recycling products from old sewage sludge ash landfills. *Journal für Kulturpflanzen* 69, 373–85.

Kratz, S., Adam, C. and Vogel, C. 2018. Pflanzenverfügbarkeit und agronomische Effizienz von klärschlammbasierten Phosphor (P)-Recyclingdüngern. In: Holm, O., Thomé-Kozmiensky, E., Quicker, P. and Kopp-Assenmacher, S. (Eds), *Verwertung von Klärschlamm*. Tagungsband zur Berliner Klärschlammkonferenz 5./6. November 2018, pp. 391–407. ISBN 978-3-944310-43-5.

Kratz, S., Vogel, C. and Adam, C. 2019. Agronomic performance of P recycling fertilizers and methods to predict it: a review. *Nutrient Cycling in Agroecosystems*. doi: https://doi.org/10.1007/s10705-019-10010-7.

Kraus, F. and Seis, W. 2015. Quantitative risk assessment of potential hazards for humans and the environment: quantification of potential hazards resulting from agricultural use of the manufactured fertilizers. Deliverable D 9.1 of the EU P-REX Project. Available at: https://zenodo.org/record/242550#.XEqyLTNCe70 (accessed on 24 January 2019).

Kraus, F., Zamzow, M. and Conzelmann, L. 2018. Ökobilanzieller Vergleich der konventionellen P-Düngemittelproduktion aus Rohphosphat mit der Phosphorrückgewinnung aus dem Abwasserpfad. In: Holm, O., Thomé-Kozmiensky, E., Quicker, P. and Kopp-Assenmacher, S. (Eds), *Verwertung von Klärschlamm*. Tagungsband zur Berliner Klärschlammkonferenz 5./6. November 2018, pp. 535–50. ISBN 978-3-944310-43-5.

Kraus, F., Zamzow, M., Conzelmann, L., Remy, C., Kleyböcker, A., Seis, W., Miehe, U., Hermann, L., Hermann, R. and Kabbe, C. 2019. Ökobilanzieller Vergleich der P-Rückgewinnung aus dem Abwasserstrom mit der Düngemittelproduktion aus Rohphosphaten unter Einbeziehung von Umweltfolgeschäden und deren Vermeidung. Kompetenzzentrum Wasser gGmbH, Proman Management GmbH, Umweltbundesamt. UBA-Texte 13/2019. Available at: https://www.umweltbundesamt.de/sites/default/files/medien/1410/publikationen/2019-02-19_texte_13-2019_phorwaerts.pdf (accessed on 7 March 2019).

Krüger, O. and Adam, C. 2014. Monitoring von Klärschlammmonoverbrennungsaschen hinsichtlich ihrer Zusammensetzung zur Ermittlung ihrer Rückgewinnungspotentiale und zur Erstellung von Referenzmaterial für die Überwachungsanalytik. UBA-Texte 49/2014. Available at: https://www.umweltbundesamt.de/sites/default/files/medien/378/publikationen/texte_49_2015_monitoring_von_klaerschlammverbrennungsaschen.pdf (accessed on 24 January 2019).

Krüger, O. and Adam, C. 2015. Recovery potential of German sewage sludge ash. *Waste Management* 45, 400–6. doi:10.1016/j.wasman.2015.01.025.

Krüger, O., Grabner, A. and Adam, C. 2014. Complete survey of German sewage sludge ash. *Environmental Science and Technology* 48(20), 11811–8. doi:10.1021/es502766x.

Kunhikrishnan, A., Shon, H. K., Bolan, N. S., El Saliby, I. and Vigneswaran, S. 2015. Sources, distribution, environmental fate, and ecological effects of nanomaterials in wastewater streams. *Critical Reviews in Environmental Science and Technology* 45(4), 277–318. doi:10.1080/10643389.2013.852407.

Kuo, S. 1996. Phosphorus. In: Sparks, D. L. (Ed.), *Methods of Soil Analysis. Part 3. SSSA Book Series 5*. SSSA, Madison, WI, pp. 869–919. ISBN 978-0-89118-866-7.

LABO. 2017. Bund/Länder Arbeitsgemeinschaft Bodenschutz: Hintergrundwerte für organische und anorganische Stoffe in Böden. 4. überarbeitete und ergänzte Auflage. Available at: https://www.labo-deutschland.de/documents/LABO_Fassung_HGW_Bericht_02_2017.pdf and https://www.labo-deutschland.de/documents/LABO_HGW_Anhang_02_2017.pdf (accessed on 25 January 2019).

Laghari, M., Naidu, R., Xiao, B., Hu, Z., Mirjat, M. S., Hu, M., Kandhro, M. N., Chen, Z., Guo, D., Jogi, Q., et al. 2016. Recent developments in biochar as an effective tool for agricultural soil management: a review. *Journal of the Science of Food and Agriculture* 96(15), 4840–9. doi:10.1002/jsfa.7753.

Lamastra, L., Suciu, N. A. and Trevisan, M. 2018. Sewage sludge for sustainable agriculture: contaminants' contents and potential use as fertilizer. *Chemical and Biological Technologies in Agriculture* 5(1), 10. doi:10.1186/s40538-018-0122-3.

Larsen, T. A., Peters, I., Alder, A., Eggen, R., Maurer, M. and Muncke, J. 2001. Re-engineering the toilet for sustainable wastewater management. *Environmental Science and Technology* 35(9), 192A–7A. doi:10.1021/es012328d.

Lekfeldt, J. D. S., Rex, M., Mercl, F., Kulhánek, M., Tlustoš, P., Magid, J. and de Neergaard, A. 2016. Effect of bioeffectors and recycled P-fertiliser products on the growth of spring wheat. *Chemical and Biological Technologies in Agriculture* 3(1), 22. doi:10.1186/s40538-016-0074-4.

Lemming, C., Bruun, S., Jensen, L. S. and Magid, J. 2017. Plant availability of phosphorus from dewatered sewage sludge, untreated incineration ashes, and other products recovered from a wastewater treatment system. *Journal of Plant Nutrition and Soil Science* 180(6), 779–87. doi:10.1002/jpln.201700206.

Lindberg, D., Molin, C. and Hupa, M. 2015. Thermal treatment of solid residues from WtE units: a review. *Waste Management* 37, 82–94. doi:10.1016/j.wasman.2014.12.009.

Lindholm-Letho, P. C., Ahkola, H. S. J. and Knuutinen, J. S. 2017. Procedures of determining organic trace compounds in municipal sewage sludge – a review. *Environmental Science and Pollution Research International* 24(5), 4383–412. doi:10.1007/s11356-016-8202-z.

Loos, R., Gawlik, B. M., Locoro, G., Rimaviciute, E., Contini, S. and Bidoglio, G. 2009. EU-wide survey of polar organic persistent pollutants in European river waters. *Environmental Pollution* 157(2), 561–8. doi:10.1016/j.envpol.2008.09.020.

Lou, H., Zhao, C., Yang, S., Shi, L., Wang, Y., Ren, X. and Bai, J. 2018. Quantitative evaluation of legacy phosphorus and its spatial distribution. *Journal of Environmental Management* 211, 296–305. doi:10.1016/j.jenvman.2018.01.062.

Luo, F., Song, J., Xia, W., Dong, M., Chen, M. and Soudek, P. 2014. Characterization of contaminants and evaluation of the suitability for land application of maize and sludge biochars. *Environmental Science and Pollution Research International* 21(14), 8707–17. doi:10.1007/s11356-014-2797-8.

Macdonald, L. M., Farrell, M., van Zwieten, L. V. and Krull, E. S. 2014. Plant growth response to biochar addition: an Australian soils perspective. *Biology and Fertility of Soils* 50(7), 1035–45. doi:10.1007/s00374-014-0921-z.

Mailler, R., Gasperi, J., Chebbo, G. and Rocher, V. 2014. Priority and emerging pollutants in sewage sludge and fate during sludge treatment. *Waste Management* 34(7), 1217–26. doi:10.1016/j.wasman.2014.03.028.

Malmborg, J. and Magnér, J. 2015. Pharmaceutical residues in sewage sludge: effect of sanitization and anaerobic digestion. *Journal of Environmental Management* 153, 1–10. doi:10.1016/j.jenvman.2015.01.041.

Meier, D., van de Beld, B., Bridgwater, A. V., Elliot, D. C., Oasmaa, A. and Preto, F. 2013. State-of-the-art of fast pyrolysis in IEA bioenergy member countries. *Renewable and Sustainable Energy Reviews* 20, 619–41. doi:10.1016/j.rser.2012.11.061.

Melo, W., Delarica, D., Guedes, A., Lavezzo, L., Donha, R., de Araújo, A., de Melo, G. and Macedo, F. 2018. Ten years of application of sewage sludge on tropical soil. A balance sheet on agricultural crops and environmental quality. *The Science of the Total Environment* 643, 1493–501. doi:10.1016/j.scitotenv.2018.06.254.

Meyer, G., Frossard, E., Mäder, P., Nanzer, S., Randall, D. G., Udert, K. M. and Oberson, A. 2018. Water soluble phosphate fertilizers for crops grown in calcareous soils – an outdated paradigm for recycled phosphorus fertilizers? *Plant and Soil* 424(1-2), 367-88. doi:10.1007/s11104-017-3545-x.

Miller, M. and O'Connor, G. A. 2009. The longer-term phytoavailability of biosolids-phosphorus. *Agronomy Journal* 101(4), 889-996. doi:10.2134/agronj2008.0197x.

Minnini, G., Blanch, A. R., Lucena, F. and Berselli, S. 2015. EU policy on sewage sludge utilization and perspectives on new approaches of sludge management. *Environmental Science and Pollution Research International* 22(10), 7361-74. doi:10.1007/s11356-014-3132-0.

Möller, K., Oberson, A., Bünemann, E. K., Cooper, J., Friedel, J. K., Glæsner, N., Hörtenhuber, S., Løes, A., Mäder, P., Meyer, G., et al. 2018. Improved phosphorus recycling in organic farming: navigating between constraints. *Advances in Agronomy* 147, 159-237. doi:10.1016/bs.agron.2017.10.004.

Nanzer, S., Oberson, A., Berger, L., Berset, E., Hermann, L. and Frossard, E. 2014. The plant availability of phosphorus from thermo-chemically treated sewage sludge ashes as studied by ^{33}P labeling techniques. *Plant and Soil* 377(1-2), 439-56. doi:10.1007/s11104-013-1968-6.

Neves, D., Thunman, H., Matos, A., Tarelho, L. and Gómez-Barea, A. 2011. Characterization and prediction of biomass pyrolysis products. *Progress in Energy and Combustion Science* 37(5), 611-30. doi:10.1016/j.pecs.2011.01.001.

Nguyen, T. T. N., Xu, C.-Y., Tahmasbian, I., Che, R., Xu, Z., Zhou, X., Wallace, H. M. and Bai, S. H. 2017. Effects of biochar on soil available nitrogen: a review and meta-analysis. *Geoderma* 288, 79-96. doi:10.1016/j.geoderma.2016.11.004.

Nnadozie, C. F., Kumari, S. and Bux, F. 2017. Status of pathogens, antibiotic resistance genes and antibiotic residues in wastewater treatment systems. *Reviews in Environmental Science and Bio/Technology* 16(3), 491-515. doi:10.1007/s11157-017-9438-x.

NWQMS. 2004. *Guidelines for Sewerage System Sludge (biosolids) Management*. National Water Quality Management Strategy, Agriculture and Resource Management Council of Australia and New Zealand (ARMCANZ) and Australian and New Zealand Environment Conservation Council (ANZECC), Canberra.

O'Connor, G. A., Sarkar, D., Brinton, S. R., Elliott, H. A. and Martin, F. G. 2004. Phytoavailability of biosolids phosphorus. *Journal of Environmental Quality* 33(2), 703-12. doi:10.2134/jeq2004.0703.

Øgaard, A. F. and Brod, E. 2016. Efficient phosphorus cycling in food production: predicting the phosphorus fertilization effect of sludge from chemical wastewater treatment. *Journal of Agricultural and Food Chemistry* 64(24), 4821-9. doi:10.1021/acs.jafc.5b05974.

Paz-Ferreiro, J., Nieto, A., Mendez, A., Askeland, M. P. J. and Gasco, G. 2018. Biochar from biosolids pyrolysis: a review. *International Journal of Environmental Research and Public Health* 15(5), 956. doi:10.3390/ijerph15050956.

Petrie, B., Barden, R. and Kasprzyk-Hordern, B. 2015. A review on emerging contaminants in wastewaters and the environment: current knowledge, understudied areas and recommendations for future monitoring. *Water Research* 72, 3-27. doi:10.1016/j.watres.2014.08.053.

Psenner, R., Pucsko, R. and Sager, M. 1984. Die Fraktionierung organischer und anorganischer Phosphorverbindungen von Sedimenten - Versuch einer Definition ökologisch wichtiger Fraktionen. *Archiv für Hydrobiologie. Supplement* 70, 111-55.

Quicker, P. and Weber, K. 2016. *Biokohle. Herstellung, Eigenschaften und Verwendung von Biomassekarbonisaten.* Springer Vieweg, Wiesbaden, Germany, 443pp. ISBN 978-3-658-03688-1.

Raju, S., Carbery, M., Kuttykattil, A., Senathirajah, K., Subashchandrabose, S. R., Evans, G. and Thavamani, P. 2018. Transport and fate of microplastics in wastewater treatment plants: implications to environmental health. *Reviews in Environmental Science and Bio/Technology* 17(4), 637–53. doi:10.1007/s11157-018-9480-3.

Rastetter, N., Rothhaupt, K. O. and Gerhardt, A. 2017. Ecotoxicological assessment of phosphate recyclates from sewage sludges. *Water, Air, and Soil Pollution* 228(4), 171. doi:10.1007/s11270-017-3331-7.

Rex, M., Drissen, P., Bartsch, S., Breuer, J. and Pischke, J. 2014. Aufschluss von Phosphor aus Klärschlamm- und Tiermehlaschen in flüssiger Konverterschlacke. Presentation at the *Workhop Abwasser-Phosphor-Dünger*, 29 January 2014, Berlin, Germany. Available at: https://www.umweltbundesamt.de/sites/default/files/medien/378/dokumente/rex_ksa-konverterschlacke.pdf (accessed on 24 January 2019).

Rezania, S., Park, J., Md Din, M. F., Mat Taib, S., Talaiekhozani, A., Kumar Yadav, K. and Kamyab, H. 2018. Microplastic pollution in different aquatic environments and biota: a review of recent studies. *Marine Pollution Bulletin* 133, 191–208. doi:10.1016/j.marpolbul.2018.05.022.

Rigby, H., Clarke, B. O., Pritchard, D. L., Meehan, B., Beshah, F., Smith, S. R. and Porter, N. A. 2016. A critical review of nitrogen mineralization in biosolids-amended soil, the associated fertilizer value for crop production and potential for emissions to the environment. *The Science of the Total Environment* 541, 1310–38. doi:10.1016/j.scitotenv.2015.08.089.

Rizzo, L., Manaia, C., Merlin, C., Schwartz, T., Dagot, C., Ploy, M. C., Michael, I. and Fatta-Kassinos, D. 2013. Urban wastewater treatment plants as hotspots for antibiotic resistant bacteria and genes spread into the environment: a review. *The Science of the Total Environment* 447, 345–60. doi:10.1016/j.scitotenv.2013.01.032.

Robinson, J. S., Baumann, K., Hu, Y., Hagemann, P., Kebelmann, L. and Leinweber, P. 2018. Phosphorus transformations in plant-based and bio-waste materials induced by pyrolysis. *Ambio* 47 (Suppl. 1), 73–82. doi:10.1007/s13280-017-0990-y.

Rokosch, A. and Heidecke, P. 2018. Klärschlammentsorgung in der Bundesrepublik Deutschland. Available at: https://www.umweltbundesamt.de/sites/default/files/medien/376/publikationen/2018_10_08_uba_fb_klaerschlamm_bf_low.pdf (accessed on 10 December 2018).

Römer, W. 2006. Plant availability of P from recycling products and phosphate fertilizers in a growth-chamber trial with rye seedlings. *Journal of Plant Nutrition and Soil Science* 169(6), 826–32.

Römer, W. and Samie, I. F. 2001. Einfluss eisenhaltiger Klärschlämme auf Kenngrößen der P-Verfügbarkeit in Ackerböden. *Journal of Plant Nutrition and Soil Science* 164(3), 321–8. doi:10.1002/1522-2624(200106)164:3<321::AID-JPLN321>3.0.CO;2-W.

Ronteltap, M., Maurer, M. and Gujer, W. 2007. The behavior of pharmaceuticals and heavy metals during struvite precipitation in urine. *Water Research* 41(9), 1859–68. doi:10.1016/j.watres.2007.01.026.

Ross, J. J., Zitomer, D. H., Miller, T. R., Weirich, C. A. and MacNamara, P. J. 2016. Emerging investigator series: pyrolysis removes common microconstituents triclocarban, triclosan and nonylphenol from biosolids. *Environmental Science: Water Research & Technology* 2(2), 282–9. doi:10.1039/C5EW00229J.

Schick, J. 2009. Untersuchungen zu P-Düngewirkung und Schwermetallgehalten thermochemisch behandelter Klärschlammaschen. Ph.D. dissertation. Technical University Braunschweig, Germany.

Schoumans, O. F., Bouraoui, F., Kabbe, C., Oenema, O. and van Dijk, K. C. 2015. Phosphorus management in Europe in a changing world. *Ambio* 44(S2), 180–92. doi:10.1007/s13280-014-0613-9.

Seiler, C. and Berendonk, T. U. 2012. Heavy metal driven co-selection of antibiotic resistance in soil and water bodies impacted by agriculture and aquaculture. *Frontiers in Microbiology* 3, 399. doi:10.3389/fmicb.2012.00399.

Severin, M., Breuer, J., Rex, M., Stemann, J., Adam, C., Van den Weghe, H. and Kücke, M. 2014. Phosphate fertilizer value of heat treated sewage sludge ash. *Plant, Soil and Environment* 60(12), 555–61. doi:10.17221/548/2014-PSE.

Shober, A. L., Hesterberg, D. L., Sims, J. T. and Gardner, S. 2006. Characterization of phosphorus species in biosolids and manures using XANES spectroscopy. *Journal of Environmental Quality* 35(6), 1983–93. doi:10.2134/jeq2006.0100.

Singh, R. P. and Agrawal, M. 2008. Potential benefits and risks of land application of sewage sludge. *Waste Management* 28(2), 347–58. doi:10.1016/j.wasman.2006.12.010.

Song, X. D., Xue, X. Y., Chen, D. Z., He, P. J. and Dai, X. H. 2014. Application of biochar from sewage sludge to plant cultivation: influence of pyrolysis temperature and biochar-to-soil ratio on yield and heavy metal accumulation. *Chemosphere* 109, 213–20. doi:10.1016/j.chemosphere.2014.01.070.

Spokas, K. A., Cantrell, K. B., Novak, J. M., Archer, D. W., Ippolito, J. A., Collins, H. P., Boateng, A. A., Lima, I. M., Lamb, M. C., McAloon, A. J., et al. 2012. Biochar: a synthesis of its agronomic impact beyond carbon sequestration. *Journal of Environmental Quality* 41(4), 973–89. doi:10.2134/jeq2011.0069.

Steckenmesser, D., Vogel, C., Adam, C. and Steffens, D. 2017. Effect of various types of thermochemical processing of sewage sludges on phosphorus speciation, solubility, and fertilization performance. *Waste Management* 62, 194–203. doi:10.1016/j.wasman.2017.02.019.

Subedi, R., Bertora, C., Zavattaro, L. and Grignani, C. 2017. Crop response to soil amended with biochar: expected benefits and unintended risks. *Italian Journal of Agronomy* 11, 161–73. doi:10.4081/ija.2017.794.

Sun, H., Gerecke, A. C., Giger, W. and Alder, A. C. 2011. Long-chain perfluorinated chemicals in digested sewage sludge in Switzerland. *Environmental Pollution* 159(2), 654–62. doi:10.1016/j.envpol.2010.09.020.

Sun, X., Shan, R., Li, X., Pan, J., Liu, X., Deng, R. and Song, J. 2017. Characterization of 60 types of Chinese biomass waste and resultant biochars in terms of their candidacy for soil application. *GCB Bioenergy* 9(9), 1423–35. doi:10.1111/gcbb.12435.

Sun, D., Hale, L., Kar, G., Soolanayakanahally, R. and Adl, S. 2018. Phosphorus recovery and reuse by pyrolysis: applications for agriculture and environment. *Chemosphere* 194, 682–91. doi:10.1016/j.chemosphere.2017.12.035.

Sunday, A. O., Abdullahi Alafara, B. and Godwin Oladele, O. 2012. Toxicity and speciation analysis of organotin compounds. *Chemical Speciation and Bioavailability* 24(4), 216–26. doi:10.3184/095422912X13491962881734.

Tirado, R. and Allsopp, M. 2012. Phosphorus in agriculture – problems and solutions. *Greenpeace.org*. Available at: http://www.greenpeace.to/greenpeace/wp-conten t/uploads/2012/06/Tirado-and-Allsopp-2012-Phosphorus-in-Agriculture-Technic al-Report-02-2012.pdf (accessed on 24 January 2019).

Torres-Dorante, L. O., Claassen, N., Steingrobe, B. and Olfs, H. W. 2005. Hydrolysis rates of inorganic polyphosphates in aqueous solution as well as in soils and effects on P availability. *Journal of Plant Nutrition and Soil Science* 168(3), 352–8. doi:10.1002/jpln.200420494.

Tsai, W. T., Liu, S. C., Chen, H. R., Chang, Y. M. and Tsai, Y. L. 2012. Textural and chemical properties of swine-manure-derived biochar pertinent to its potential use as a soil amendment. *Chemosphere* 89(2), 198–203. doi:10.1016/j.chemosphere.2012.05.085.

Van Dijk, K. C., Lesschen, J. P. and Oenema, O. 2016. Phosphorus flows and balances of the European Union Member States. *The Science of the Total Environment* 542(B), 1078–93. doi:10.1016/j.scitotenv.2015.08.048.

Vaneeckhaute, C., Janda, J., Vanrolleghem, P. A., Tack, F. M. G. and Meers, E. 2016. Phosphorus use efficiency of bio-based fertilizers: bioavailability and fractionation. *Pedosphere* 26(3), 310–25. doi:10.1016/S1002-0160(15)60045-5.

Verlicchi, P. and Zambello, E. 2015. Pharmaceuticals and personal care products in untreated and treated sewage sludge: occurrence and environmental risk in the case of application on soil – a critical review. *The Science of the Total Environment* 538, 750–67. doi:10.1016/j.scitotenv.2015.08.108.

Vogel, T., Nelles, M. and Eichler-Löbermann, B. 2015. Phosphorus application with recycled products from municipal waste water to different crop species. *Ecological Engineering* 83, 466–75. doi:10.1016/j.ecoleng.2015.06.044.

Vollaro, M., Galioto, F. and Viaggi, D. 2016. The circular economy and agriculture: new opportunities for re-using phosphorus as fertilizer. *Bio-Based and Applied Economics* 5, 267–85.

Wang, T., Camps-Arbestain, M., Hedley, M. and Bishop, P. 2012. Predicting phosphorus bioavailability from high-ash biochars. *Plant and Soil* 357(1–2), 173–87. doi:10.1007/s11104-012-1131-9.

Webber, M. D. and Sidhwa, P. 2007. Land application of sewage biosolids: are Canadian trace metal guidelines/regulations over-protective for crop production? In: LeBlanc, R. J., Laughton, P. J. and Yagi, R. (Eds), *Moving Forward Wastewater Biosolids Sustainability: Technical, Managerial and Public Synergy*. New Brunswick. GMSC 2007, pp. 463–70.

Weissengruber, L., Möller, K., Puschenreiter, M. and Friedel, J. K. 2018. Long-term soil accumulation of potentially toxic elements and selected organic pollutants through application of recycled phosphorus fertilizers for organic farming conditions. *Nutrient Cycling in Agroecosystems* 110(3), 427–49. doi:10.1007/s10705-018-9907-9.

Wiechmann, B., Dienemann, C., Kabbe, C., Brandt, S., Vogel, I. and Rokosch, A. 2013. *Sewage Sludge Management in Germany*. Umweltbundesamt (UBA), Germany. Download from. Available at: https://www.umweltbundesamt.de/sites/default/files/medien/378/publikationen/sewage_sludge_management_in_germany.pdf (accessed on 24 January 2019).

Wilken, V. and Kabbe, C. 2015. Sustainable sewage sludge management fostering phosphorus recovery and energy efficiency (EU-Project P-REX). Deliverable D 8.1 Quantification of nutritional value and toxic effects of each P recovery product. Available at: https://zenodo.org/record/242550#.XEqyLTNCe70 (accessed on 24 January 2019).

Wollmann, I., Gauro, A., Müller, T. and Möller, K. 2018. Phosphorus bioavailability of sewage sludge-based recycled fertilizers. *Journal of Plant Nutrition and Soil Science* 181(2), 158-66. doi:10.1002/jpln.201700111.

Xu, H., Zhang, H., Shao, L. and He, P. 2012. Fraction distributions of phosphorus in sewage sludge and sludge ash. *Waste and Biomass Valorization* 3(3), 355-61. doi:10.1007/s12649-011-9103-5.

Xu, X., Cao, X., Zhao, L. and Sun, T. 2014. Comparison of sewage sludge- and pig manure-derived biochars for hydrogen sulfide removal. *Chemosphere* 111, 296-303. doi:10.1016/j.chemosphere.2014.04.014.

Yuan, H., Lu, T., Wang, Y., Chen, Y. and Lei, T. 2016. Sewage sludge biochar: nutrient composition and its effect on the leaching of soil nutrients. *Geoderma* 267, 17-23. doi:10.1016/j.geoderma.2015.12.020.

Zeggel, L., Riedel, H. and Marb, C. 2015. Rückholbarkeit von Phosphor aus kommunalen Klärschlämmen - Abschlussbericht; Bayerisches Landesamt für Umwelt. Available at: https://www.bestellen.bayern.de/application/eshop_app000005?SID=149040 3870&ACTIONxSESSxSACTION(BILDxKEY:%27lfu_abfall_00221%27,BILDxCLA SS:%27Artikel%27,BILDxTYPE:%27PDF%27) (accessed on 24 January 2019).

Zennegg, M., Munoz, M., Schmid, P. and Gerecke, A. C. 2013. Temporal trends of persistent organic pollutants in digested sewage sludge (1993-2012). *Environment International* 60, 202-8. doi:10.1016/j.envint.2013.08.020.

Zhang, Q., Liu, H., Li, W., Xu, J. and Liang, Q. 2012. Behaviour of phosphorus during co-gasification of sewage sludge and coal. *Energy and Fuels* 26(5), 2830-6. doi:10.1021/ef300006d.

Zhang, H. L., Fang, W., Wang, Y. P., Sheng, G. P., Zeng, R. J., Li, W. W. and Yu, H. Q. 2013. Phosphorus removal in an enhanced biological phosphorus removal process: role of extracellular polymeric substances. *Environmental Science and Technology* 47(20), 11482-9. doi:10.1021/es403227p.

Zielińska, A., Oleszczuk, P., Charmas, B., Skubiszewska-Zięba, J. and Pasieczna-Patkowska, S. 2015. Effect of sewage sludge properties on the biochar characteristic. *Journal of Analytical and Applied Pyrolysis* 112, 201-13.

Chapter 3

Safe and sustainable use of bio-based fertilizers in agricultural production systems

April Leytem*, Robert Dungan, Mindy Spiehs and Dan Miller, United States Department of Agriculture, USA

1 Introduction

Development of circular agricultural production systems necessitates the reuse/recycling/upcycling of agricultural byproducts (e.g. animal manure, crop residues, food waste and food processing waste) back into the production cycle thereby lessening the overall footprint of agriculture. Historically, many agricultural byproducts have been classified as 'wastes', as they were not the primary product of production. However, the goal of circularity is to utilize these 'wastes' as a resource instead of simply discarding them. Many agricultural byproducts contain significant amounts of nitrogen (N), phosphorus (P) and potassium (K) as well as a wide variety of micronutrients which makes them an excellent nutrient source for crop growth. Because of this, the development of bio-based fertilizers from a variety of agricultural byproducts is of increasing interest. In addition to supplying nutrients, bio-based fertilizers can improve soil health via the addition of organic carbon (C) which may improve soil structure, water holding capacity and water infiltration and support a healthy microbiome. Although these products contain valuable N and P for agricultural production, the conversion efficiency into edible food or other end products

* Corresponding author: april.leytem@usda.gov, 01-208-423-6530, USDA-ARS NWISRL, 3793N 3600E, Kimberly, ID, 83341, USA.

http://dx.doi.org/10.19103/AS.2023.0120.16
Published by Burleigh Dodds Science Publishing Limited, 2024.

tends to be low with typical use efficiencies of 30% or less (Bordirsky et al., 2012; UNEP and WHRC, 2007; FAO, 2006) leading to potential environmental losses that impact air and water quality as well contribute greenhouse gasses (GHG) to the atmosphere which are associated with climate change. Efficiently integrating these bio-based fertilizers back into agricultural production is dependent on the characteristics of the product, availability, quality control, transportation costs/logistics, environmental regulation and public acceptance (Westerman and Bicudo, 2005). Additionally, valorization of these products is challenging as the nutrient contents tend to be diluted compared to synthetic fertilizer, less consistent and in some cases harder to handle and apply than traditional fertilizers. However, the additional benefits of providing a wider variety of micronutrients, compared to synthetic fertilizer, along with additions of organic C that may improve overall soil health and productivity, need to be taken into consideration as well.

In addition to concerns related to nutrient pollution, other environmental contaminants that may be associated with bio-based fertilizers include heavy metals, pathogens, antibiotics and chemicals of emerging concern (CEC; e.g. pesticides, hormones, polychlorinated biphenyls [PCBs], polycyclic aromatic hydrocarbons [PAHs] and microplastics; Tran et al., 2018; O'Connor et al., 2022; Tian et al., 2022). Bio-based fertilizers originating from biosolids, manures and other organic wastes can contain heavy metals (O'Connor et al., 2021). Heavy metals are toxic in high concentrations and are persistent pollutants in the environment where they can bioaccumulate and biomagnify through the food chain (Sardar et al., 2013). Pathogens may be present in many organic byproducts (livestock manure, manure compost, biosolids) used as bio-based fertilizers. The application of these fertilizers to agricultural crops, in particular those grown for fresh produce, has resulted in foodborne illness outbreaks (Alegbeleye et al., 2018; Callejón et al., 2015). The use of antibiotics in animal production and the presence of antibiotics in biosolids have raised concerns about the release of these drugs into the environment, particularly with land application of these byproducts, and the increasing prevalence of antibiotic resistant bacteria (ARB) and antibiotic resistance genes (ARGs) in the environment. Pesticide residues are commonly found in food wastes which may pose a threat to the quality of composts and digestates generated from these byproducts (Nguyen et al., 2020). The persistence and phytotoxic effects of these residues are unknown but may cause damage to soil quality and plant growth when applied to cropland as bio-based fertilizers at high rates (Boudh and Singh, 2019). There are a host of other CEC that are introduced into agricultural production and they remain persistent within the system. These chemicals have the potential to damage soil health and can pose health risks to both animals and humans. Given the large variety of agricultural byproducts that can serve as bio-based

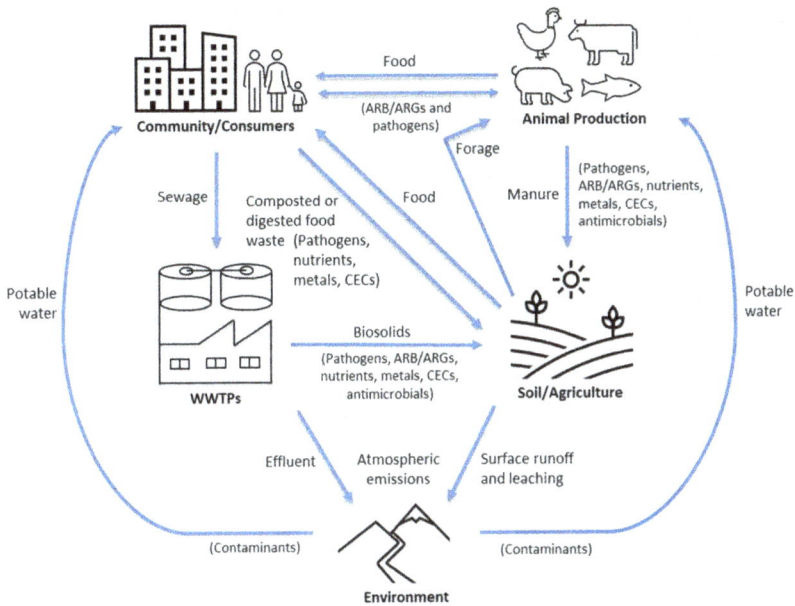

Figure 1 The routes by which biological and chemical contaminants in bio-based fertilizers are spread in environmental ecosystems. The contaminants make their way into the environment when manure, food waste and biosolids are applied to soils, either through direct application, animal excretion or disposal. The contaminants can then be transported in surface runoff or by leaching to surface and ground water, respectively, or emitted to the atmosphere. Humans and animals can be continuously exposed to contaminants via ingestion of food and water, while humans can also be exposed via contact in recreational waters.

fertilizers available for recycling within the agricultural system, it is important not only to consider the benefits in terms of C and nutrients, but also to assess any potential tradeoffs with respect to contaminants. Figure 1 illustrates some of the potential pathways of contaminant flow within circular systems.

2 Risk factors associated with utilizing bio-based fertilizers in agricultural production

2.1 Nutrient losses to the environment

Estimates of the conversion of N used in food production to that consumed by humans range from 10% to 20% (Bordirsky et al., 2012; UNEP and WHRC, 2007), with greater than 50% of N applied to cropland lost to downstream and downwind environments (Davidson et al., 2012). It is estimated that only 15% to 30% of applied phosphorus (P) is taken up by harvested crops (FAO, 2006). With these low nutrient use efficiency rates, the integration of agricultural byproducts as bio-based fertilizers into agricultural production needs to be carefully managed to reduce nutrient losses to the environment.

In terms of nutrients generated, livestock manure is one of the largest sources of bio-based fertilizers available. Globally, it is estimated that livestock production generates 131 and 23 Tg of manure N and P, respectively, per year (Liu et al., 2017; Zhang, 2017). Manure nutrient production is expected to increase as human and livestock populations increase and the dietary intake of meat increases in developing countries (Herrero and Thornton, 2013). While the quantity of manure nutrients generated over time are increasing, the growing disconnect between livestock and crop production on the agricultural landscape has left most modern livestock producers with a surplus of on-farm nutrients which contribute to environmental degradation in many regions. Overall, the greatest concerns have been related to air quality degradation due to emissions of NH_3, water quality impacts from nitrate (NO_3) leaching and P losses via runoff and climate impacts and ozone depletion from nitrous oxide (N_2O) emissions (Holly et al., 2018).

2.1.1 Reactive nitrogen losses to the environment

Reactive N (N_r) is essential to the growth of plants and animals and is typically the most limiting nutrient in agricultural production. The current rate of N_r loss to the environment, due to agricultural production, is more than 10 times the rate that occurred at the end of the 1800s (UNEP and WHRC, 2007) largely due to industrial-scale plant fertilizer production using the Haber-Bosch process in the twentieth century. Nitrogen losses are of increasing concern due to their negative consequences on human, animal and environmental health. The primary forms of N_r loss are ammonia (NH_3), nitrate (NO_3), nitrous oxide (N_2O) and N oxides (NO and NO_2 or NO_x).

Food production, including cropland and livestock production, is estimated to generate up to 90% of global NH_3 emissions (Ma et al., 2021; Zhan et al., 2021). Liu et al. (2022) estimated that, in 2010, approximately 58 Tg of N was lost as NH_3, with approximately 67% of that being generated from livestock production and land application of manures. Ammonia in the atmosphere is an important driver for the formation of fine particulate matter ($PM_{2.5}$) which is both an air quality and human health concern. In addition, the deposition of this N into terrestrial ecosystems can cause soil acidification, lead to eutrophication of sensitive waterbodies, affect biodiversity, lead to secondary N_2O emissions and cause changes in terrestrial carbon sinks (Liu et al., 2019; Elser et al., 2009; Zhan et al., 2017; Wang et al., 2017; Sutton et al., 2014). Liu et al. (2022) reported that global agricultural NH_3 emissions and N deposition increased by 78% and 72% from 1980 to 2018, respectively. Understanding the sources of NH_3 emissions and temporal and spatial patterns of N deposition will be essential for developing abatement strategies.

Reductions in NH_3 emissions from livestock via improved feed efficiency and enhancing NH_3-N capture from livestock manure and recycling that N back into cropping systems can help improve the overall N use efficiency and circularity of agricultural systems. In addition, development of enhanced efficiency bio-based fertilizers that are stable and release N slowly, as well as improved N management (better matching N to crop needs over time, immediate incorporation of fertilizers into soils, etc.), can also help minimize NH_3 volatilization losses in cropping systems.

Nitrate is one of the most widespread groundwater contaminants in the world, and its consumption has been linked with a variety of human health ailments. The main sources of NO_3 pollution of groundwater are from intensive use of N fertilizers (both synthetic and livestock manures) in agricultural production, disposal of sanitary and industrial wastes, leakage from septic systems and landfills and NO_x air stripping waste from air pollution control devices (Bhatnagar and Sillanpää, 2011; Abascal et al., 2022). Transport of NO_3 from ground waters to surface waters as well as runoff of NO_3 to surface waters has led to eutrophication in N limited systems, resulting in hypoxia and anoxia, loss of biodiversity and harmful algal blooms damaging fisheries and pristine marine environments (BijaySingh and Craswell, 2021). Optimizing N application rates and timing can be effective approaches to reducing NO_3 losses to ground and surface waters. In addition, development of more stable bio-based fertilizers with slow N release characteristics can reduce losses by better matching N release with plant uptake.

Nitrous oxide is a potent GHG as well as a stratospheric ozone depleting substance with a lifetime of 116 years (Prather et al., 2015). The largest global source of anthropogenic N_2O emissions is agriculture, representing 52% of total emissions (Tian et al., 2020). Tian et al. (2020) reported that the largest increases in anthropogenic N_2O emissions in the last four decades were due to direct emissions from agricultural soils receiving N additions (71%). There is also considerable positive feedback between climate change and increasing N_2O emissions. As the recent growth in N_2O emissions have exceeded some of the highest projected emission scenarios, mitigating these emissions is urgent (Gidden et al., 2019). Reducing excess N application to croplands and adopting precision fertilizer application methods provide the greatest immediate opportunities for the abatement of N_2O emissions.

Improving the connectivity between agricultural byproducts and agricultural production can improve N use efficiency and reduce N_r losses through more efficient recycling of nutrients. Application of bio-based fertilizers following the 4R strategy of applying the right fertilizer at the right rate and time with the right application method can enhance overall system efficiencies and thereby reduce their environmental impact.

2.1.2 Phosphorus losses

Globally, P deficiency is considered a major limitation for crop production, particularly in low-input agricultural systems (Raghothama, 2005). However, in many developed countries, over application of P as fertilizer (synthetic and bio-based fertilizers) has become a significant source for water pollution. In areas with concentrated livestock production, managing manure P to reduce negative environmental impacts is a challenge. Although manure P availability is greater or similar to synthetic fertilizer P use, manure P is not efficiently used in crop production. The inefficient use of manure P is due to several factors: uneven distribution of manure by grazing animals, incomplete collection and inadequate storage of manure from housed animals, poor timing of manure application, high cost of transporting manure, manure N:P ratios that do not necessarily match N:P needs of crops resulting in overapplication of P in some instances and the relatively low price of synthetic P fertilizer. Due to the high moisture content and bulky nature of manures, they are generally land applied within close proximity of where they are produced which had resulted in the buildup of P in soils surrounding livestock farms. This excessive P application has led to P surpluses in croplands, decreased P use efficiency and increasing P losses to surface waters.

Elevated P concentrations in receiving waters can lead to eutrophication, which can be costly to remediate. For example, in England and Wales, it has been estimated that damages due to agricultural losses of P are near £19 million (Bateman et al., 2011). In some countries, direct discharge of P in wastewater to surface waters is still common. For example, in Thailand, P containing wastewaters discharged directly to surface waters from dairy and swine farms were estimated to add 554 and 261 T P y^{-1} (Prathumchai et al., 2018). These discharges have a direct negative impact on surface water quality in these regions. Phosphorus surplus has been shown to increase with increasing livestock density in studies at continental, national and regional scale, especially at livestock densities above 2 LSU ha^{-1} (Liu et al., 2017; Nesme et al., 2015). One of the main causes of these surpluses is the large amount of P imported in feed coupled with low P use efficiency of most livestock. Therefore, there is often a clear relationship between livestock density and P balances at the farm level. In several countries (e.g. the United States, Netherlands, Norway, Denmark, Finland), manure P generated can meet or even exceed that needed for sustainable crop production in some regions (Smit et al., 2015; Hanserud et al., 2016; Yang et al., 2016; Parchomenko and Borsky, 2018; Svanbäck et al., 2019). Despite this large potential for within country P recycling, areas with the largest amounts of manure P are not necessarily co-located with the highest P needs. Therefore, hot spots occur where manure P exceeds that regionally needed for crop production. In addition, fertilizer P is still often used in these production systems leading to a further buildup of soil P.

Development of technologies to capture and concentrate P from agricultural byproduct streams will be essential to enable more sustainable recycling of P back through the production system. Additionally, cost-effective methods for redistribution of byproduct P from areas of surplus to areas of deficit will need to be developed. In some instances, particularly in livestock dense regions, regulations, economic incentives and technical solutions for enhanced relocation of byproduct P from areas with surplus to areas with deficit will be crucial. Support for processing and trading of byproduct-based P fertilizers can help reduce nutrient imbalances between areas with excess nutrient and those needing this valuable nutrient.

2.2 Heavy metals

Heavy metal contamination of byproducts used as bio-based fertilizers is a concern as these metals can bioaccumulate in agricultural crops and subsequently endanger the health of the consumer (Polechónska et al., 2018). Two of the most common byproducts used as bio-based fertilizers are manures and sewage sludges which may be applied raw or composted. On average, sludges have the highest values of heavy metal concentrations, compared to composts and manure, particularly when wastewater treatment plants also collect industrial effluents (Lopes et al., 2011). Cajamarca et al. (2019) characterized several agro-industrial, livestock and food wastes used to create bio-based fertilizers and found that some material contained high levels of trace metals, in particular cadmium (Cd). Livestock manures can contain elevated levels of arsenic (As), copper (Cu) and zinc (Zn) as these are found in feed additives to improve overall health and growth of livestock (Bloem et al., 2017). Repeated application of these bio-based fertilizers to cropland can lead to buildup of metals in soils which can potentially enter the food chain. Monte Carlo simulations indicated that repeated application of bio-based fertilizer likely increased the concentrations of Zn, Cd and As in soil compared with soil background levels (Gong et al., 2019). Concern over heavy metal contamination is ubiquitous where sludges, manures, composts and other bio-based fertilizers are applied to agricultural lands resulting in legislation in many countries regulating application (Lopes et al., 2011; Shi et al., 2018; Gong et al., 2019; Nunes et al., 2021).

In China, it was reported that livestock manure was one of the predominant sources of trace metals entering agricultural soils, accounting for approximately 55%, 69% and 51% of the total Cd, Cu and Zn inputs, respectively (Luo et al.,

2009; Shi et al., 2018). Tan and Tran (2021) reported that Cu and Zn accumulated in soil, water and rice with application of sewage sludge at varying levels with concentrations exceeding permissible Vietnamese standards (QCVN 03: 2008) and US EPA 503 criteria. A risk assessment focused on reuse of organic waste as bio-based fertilizers, indicated that Zn was the main contributor to total risk due to its high concentration in bio-based fertilizers and high bio-transfer potential (Lopes et al., 2011). While others reported that health risk index values for Cd, cobalt (Co) and lead (Pb) suggested that these metals had the probability to cause health problems in people who consume vegetables grown with bio-based fertilizers (Ugulu et al., 2021). Others have reported heavy metal contents of bio-based fertilizers being below regulatory thresholds (Nekvapil et al., 2021; Golovko et al., 2022). In addition to health concerns, elevated levels of heavy metals in agricultural soils can result in decreased germination, altered metabolism, growth reduction, reduced biomass production and reduced yield for sensitive plants (Sethy and Ghosh, 2013; Goyal et al., 2020). While phytoremediation of metals from soils is feasible, there are few remediation strategies available, therefore prevention is key.

2.3 Pathogens

Pathogens can be present in food wastes, manures and other byproducts that are used as bio-based fertilizers. This contamination can pose potentially serious risks to human health as the application of these bio-based fertilizers to croplands can result in foodborne and waterborne outbreaks. The risks will vary from low to high, depending upon a number of factors, such as animal health, microbial concentration, land application method and environmental conditions (Venglovsky et al., 2009; Adegoke et al., 2016). Some pathogens commonly found in manures and food wastes are *Campylobacter*, *Escherichia coli* O157:H7, *Clostridium*, *Listeria*, *Salmonella*, hepatitis E virus, *Cryptosporidium parvum*, *Giardia lamblia* and *Shigella* (Kraus et al., 2003; Pepper et al., 2006; Manyi-Loh et al., 2016; O'Connor et al., 2022). The level of these zoonotic pathogens can exceed thousands per gram of material, with infection causing temporary illness or mortality, especially in high-risk individuals (Hutchison et al., 2005; Klein et al., 2010; Létourneau et al., 2010). Exposure of humans to pathogens can occur through occupational and recreational exposures, ingestion of contaminated food and water or aerogenic routes (Matthews, 2006; Dungan, 2010). According to the Centers for Disease Control and Prevention (CDC), *Escherichia coli* O157:H7, *Listeria monocytogenes* and *Salmonella* spp. are three major foodborne pathogens (CDC, 2020). These bacteria have been associated with numerous outbreaks in the United States (Ratnam et al., 1988; Nightingale et al., 2004; CDC, 2020).

Application of contaminated bio-based fertilizers to soils, particularly surface application, can result in the transport of pathogens to surface or ground waters (Abu-Ashour et al., 1994; Jamieson et al., 2002; Tyrrel and Quinton, 2003; Bloem et al., 2017). The overland transport of microorganisms is called horizontal movement, while the leaching of microorganisms through soil and other porous subsurface strata is referred to as vertical movement. Unless a soil is saturated or contains an impermeable barrier, vertical movement of microorganisms will occur (Mawdsley et al., 1995). Some physical and chemical properties that influence the vertical movement of microorganisms are soil type, water content and water flow, microbe and soil particle surface properties, cell motility, pH, plant roots, temperature and presence of micro- and meso-faunal organisms (Mawdsley et al., 1995; Unc and Goss, 2004).

Rapid horizontal transport of microorganisms to surface waters can occur when either the rainfall intensity exceeds the soil's infiltration rate or when the soil becomes so saturated that no rainfall can percolate (Tyrrel and Quinton, 2003). Factors that influence the level of microbiological contamination in runoff from agricultural lands are organism die-off rates, quantity and type of amendment applied, sloping terrain, rainfall intensity and water infiltration rate (Evans and Owens, 1972; Doran and Linn, 1979; Baxter-Potter and Gilliland, 1988; Abu-Ashour and Lee, 2000; Jenkins et al., 2006; Ramos et al., 2006). Methods to mitigate the offsite transport of microorganisms in runoff from amended soils include use of vegetative filter strips (Coyne et al., 1995; Fajardo et al., 2001) or vegetative treatment systems with a settling basin for solids collection and a vegetated area (Koelsch et al., 2006; Berry et al., 2007).

Alternatively, bio-based fertilizers can be treated prior to land application, thus reducing subsequent risks associated with pathogens (Lund et al., 1996; Tiquia et al., 1998). Various physical, chemical and biological treatment technologies could be used to reduce or eliminate the presence of pathogens (Heinonen-Tanski et al., 2006). While there are advantages and disadvantages with these methods, some can provide additional benefits, such as the production of compost that can be used to enhance the properties of agricultural soils (Tester, 1990) or biogas for energy generation (Holm-Nielsen et al., 2009). There are a wide variety of technologies available to treat byproducts; however, the only processes with a documented record of cost-effective pathogen reduction are composting and anaerobic digestion (Sobsey et al., 2006; Martens and Böhm, 2009; Gurtler et al., 2018; Jiang et al., 2020; Thakalie and MacRae, 2021).

2.4 Antibiotics, antibiotic resistant bacteria and antibiotic resistance genes

In many countries, antibiotics are used therapeutically (high doses) in livestock production to treat specific diseases or subtherapeutically (low doses) by incorporating into feed to improve growth efficiency (Sarmah et al., 2006). It is also common practice to simultaneously administer multiple classes of antibiotics to livestock at the production facility (Song et al., 2007). Because not all antibiotics are absorbed in the gut of animals, they are excreted via urine and feces in unaltered form and as metabolites (Halling-Sørensen et al., 1998; Boxall et al., 2004). It has been estimated that as much as 80% of orally ingested antibiotics can be excreted in urine and feces (Elmund et al., 1971; Levy, 1992; Halling-Sørensen et al., 2002). Several classes of veterinary pharmaceuticals and antibiotics, including coccidiostats, ionophores, lincosamides, macrolides, sulfonamides and tetracyclines, have been detected in surface waters adjacent to livestock operations (Campagnolo et al., 2002; Hao et al., 2006; Song et al., 2007). In addition, the practice of land applying livestock manure as a bio-based fertilizer provides for the introduction of antibiotics over large areas in the environment, resulting in frequent detection of these compounds in soils and waters worldwide (Hamscher et al., 2002; Christian et al., 2003; Kemper, 2008).

Antibiotics can select for ARB in the animal gut, which are then released into the environment via excreted feces. ARGs are the genetic code ARB use to produce proteins responsible for antibiotic resistance. ARGs can be distributed to similar, distantly related and pathogenic bacteria through horizontal gene transfer mechanisms (Alekshun and Levy, 2007) and are considered an emerging contaminant (Pruden et al., 2006). Prophylactic and therapeutic uses of antibiotics in food–animal production has the potential to contribute to drug resistant bacteria in vivo, during manure storage, and in soils receiving manure solids or wastewater as a fertilizer (Binh et al., 2008; Negreanu et al., 2012; Chantziaras et al., 2014). Use of antimicrobials in livestock production may also intensify the resistance of pathogens to antibiotics, reducing the ability to treat infected individuals (Boxall et al., 2003; Bahe et al., 2006).

The continued use of large quantities of antibiotics in animal production and human health raises concerns about the release of these drugs and the increasing prevalence of ARB and ARGs in the environment when livestock manure and municipal biosolids are used as bio-based fertilizers. While use of antibiotic drugs can enrich ARB, the interplay between antibiotic drug use, ARB/ARGs and land use practices in agroecosystems is poorly understood (Williams-Nguyen et al., 2016). Manure application can transfer ARB and ARGs to soils, as well as antibiotic residues and other xenobiotic compounds, resulting in the expansion of antibiotic resistance reservoirs when compared to that of native soils (Heuer and Smalla, 2007; Cytryn, 2013; Amarakoon et al., 2016; Dungan et al., 2019). The detection of ARGs in soils, manures and agriculturally impacted environments is well documented in the scientific literature, but the risk of elevated ARG levels on public health is not well understood. Low concentrations of antibiotics and their metabolites can enter the food chain when plants are grown on fields which have received bio-based fertilizers contaminated by antibiotics (Bloem et al., 2017). This could potentially contribute to antibiotic resistance in humans and animals when edible crops are contaminated by traces of antibiotics, as it has been shown that even low-level antibiotic concentrations can select for antibiotic resistance in bacteria (Sandegren, 2014).

To minimize the effects of antibiotics, ARB and ARG in bio-based fertilizers, removal of these contaminants prior to land application may be necessary. Removal of antibiotics during anaerobic digestion has had varying success, even within the same class of antibiotics, while composting has been shown to significantly reduce antibiotics in nearly all cases (Youngquist et al., 2016). The removal of ARB and ARGs with digestion and composting varies with treatment effectiveness dependent on temperature (Beneragama, et al., 2013; Diehl and La Para, 2010; Guan et al., 2007; Resende et al., 2014). Alternatively, bioremediation using algae (Guo and Chen, 2015; Yu et al., 2017; Waseem et al., 2017) and fungal (Naghdi et al., 2018) species have successfully removed antimicrobials in water treatment facilities.

2.5 Chemicals of emerging concern

The trace levels of emerging or recently detected pollutants in soil and water receiving bio-based fertilizers is a growing concern for human health and the environment because at high concentrations, these chemicals can affect animal, human and environmental health. This category of emerging contaminants consists of pharmaceuticals, pesticides, hormones, disinfectants and their metabolites (Chaturvedi et al., 2021; Rathi et al., 2021). Hormones may be naturally occurring endogenous hormones, or natural or synthetic exogenous hormones used as growth promoters. Hormones found in animal waste are a concern as they may act as endocrine disruptors and influence the production of natural hormones, providing the blockage, minimization, stimulation or their inhibition (Matthiessen et al., 2002). Hormones have been found in high concentrations in pig farming effluent (Honorio et al., 2019; Fine et al., 2003). Hutchins et al. (2007) reported estrogens and estrogen metabolites in swine sow, finisher and nursery lagoons, as well as dairy, beef and poultry lagoons. Some hormones have even been detected in groundwater near livestock wastewater lagoons (Fine et al., 2003; Bartelt-Hunt et al., 2011).

Pesticides and microplastics are commonly found in food wastes due to their presence in soils and on foods (Nguyen et al., 2020; Tian et al., 2022). Some pesticides adsorb strongly to soil particles and can accumulate in soils and potentially be taken up by plants (Li et al., 2018). Pesticide residues that persist in bio-based fertilizers can damage soil quality as well as plant health

(Boudh and Singh, 2019). Composts containing residual herbicides were found to inhibit seedling vigor, emergence rate and plant growth in sensitive plants when mixed with soil at rates of 10–20% (Chang et al., 2017). Some plant growth regulator herbicides do not break down easily and can pass through animal digestive systems and composting facilities and have the potential to cause significant damage to plants and crops.

A wide range of microplastics composed of polyethylene, polypropylene, polyamide, polyester, polystyrene, polyethylene terephthalate, polyvinyl chloride, polyurethane, polyvinylidene chloride and polyacrylonitrile have been identified in the environment. Polymer-coated, slow-release fertilizers, plastic mulching for weed and pathogen control and other cultivation practices are important sources contributing to microplastics in soil. Composting, which usually reduces CEC in manures and agricultural wastes, accelerates microplastic formation when larger plastic contaminants become more fragmented. Although ingestion by earthworms and other soil fauna would be the most direct biological effect, microplastic's capacity to serve as particles capable of concentrating other contaminants, such as antibiotics and pharmaceutically active compounds, may be an unrecognized contributor to CEC risk.

PCBs are a widely recognized environmental and food contaminant, with human exposure primarily coming from consumption of animal-derived food such as meat, dairy and eggs, and fishery products (Weber et al., 2018). Animal feeds and feed additives are major sources of dioxins and PCB contamination for food of animal origin. Animals that are exposed to contaminated soil, such as beef cattle and veal from suckler cow herds, sheep and free-range chickens, have the highest risk of exposure through uptake of contaminated soil, grass, silage and hay. Application of sewage sludge and contaminated sediment deposits on agricultural land caused elevated PCB levels in meat from sheep and beef cattle grazing the contaminated areas (Weber et al., 2018). Management recommendations include not grazing contaminated areas, feeding non-contaminated feed sources during the finishing phase prior to slaughter, increasing the cutting height of plants grown on contaminated soil to reduce exposure to contaminated soil and mixing non-contaminated feed with contaminated feed sources to dilute the PCB in the whole feed. PAHs are classified as carcinogenic, mutagenic, teratogenic and immunotoxic to micro- and macro-organisms (Patel et al., 2020). PAHs are commonly found in aquatic and terrestrial environments and are strongly adsorbed to soil particles where they can enter the food chain due to plant uptake of contaminated soils (O'Connor et al., 2022).

Assuming increasing use of bio-based fertilizers for food production in the future, it would be beneficial to have concentration limits for CEC. Risk estimation based on risk quotients (RQs) indicated generally low environmental risks associated with application of bio-based fertilizer to soils for food crop production. However, the toxicity and potential concentration and build-up in soils of CEC mixtures needs to be considered when estimating the risks from application of bio-based fertilizers on agricultural land or in other production systems.

3 Case study: intensive dairy production in the northwest United States

3.1 Background

Intensive livestock production regions generate large volumes of manure that can be recycled back through crops, creating more sustainable circular systems. However, the large volume of manure and the potential contaminants found within can also create challenges. As an example, we will discuss several aspects of an intensive dairy production region and issues regarding sustainability/ circularity of using manure as a bio-based fertilizer within the region. Idaho is the third largest dairy producing state in the United States with 70% of the lactating herd or ~894 000 animals (including mature and young stock) located in South Central Idaho, also known as the 'Magic Valley' region (Fig. 2). This region has approximately 429 000 ha of cropland resulting in an animal density of ~2 AU ha^{-1} which has been shown to be the stocking rate at which regional nutrient surpluses can occur (Liu et al., 2017; Nesme et al., 2015). All the manure generated within the region is land applied, with a majority applied either as a solid (some of which is composted) or a liquid (following solid separation) which is blended with irrigation water and applied through the irrigation system.

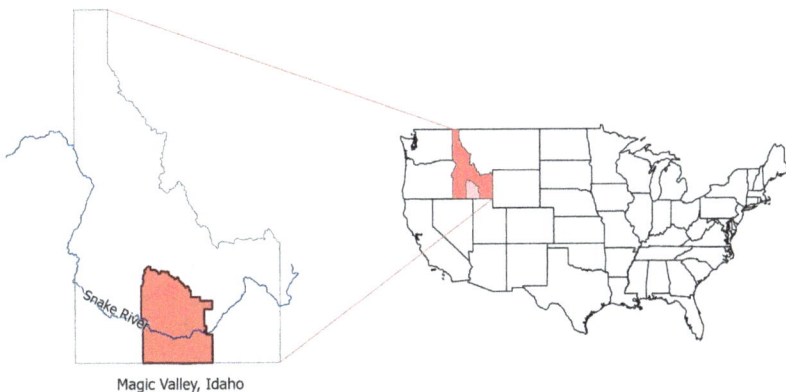

Magic Valley, Idaho

Figure 2 The Magic Valley region of South Central Idaho. This region is the third largest dairy producing region in the United States with ~894 000 animals (including mature and young stock) and approximately 429 000 ha of cropland resulting in an animal density of ~2 AU ha^{-1}.

This region has a semi-arid climate with an annual average precipitation of only 270 mm, which mainly occurs during the non-growing season. Therefore, all cropland in the region is irrigated with water diverted from the Snake River. This water is managed by the Twin Falls Canal Company (TFCC) south of the Snake River and the Northside Canal Company to the north. The TFCC routes irrigation water through 180 km of main canals and over 1600 km of smaller channels and laterals. Irrigation water flows by gravity from the Snake River throughout the 82 000 ha watershed (Fig. 3). Natural channels or coulees often convey water to laterals and collect runoff and return flow from fields. These return flows ultimately convey water back to the Snake River and can carry nutrients and other contaminants that may impair water quality. This irrigation tract has served as a benchmark watershed for the USDA Agricultural Research Service since 2004 and provides a unique system in which to evaluate the impact of land management practices on water quality in the region. In this section, we discuss the impacts of the recycling of manure products on nutrient balances and losses as well as the impact on antibiotic resistance in the regional environment.

Figure 3 The Twin Falls Canal Company irrigation system including sampling sites for studies located within the watershed. The sampling sites within the Upper Snake Rock Watershed in South Central Idaho: N coulee (NC), Deep Creek (DC), Mud Creek (MC), I Coulee (IC), Cedar Draw (CD), Rock Creek Poleline (RCP), Twin Falls Coulee (TFC), Hansen Coulee (HC), Milner Dam (MD) and Rock Creek (RC). Both MD and RC were upstream sites and used to determine background concentrations. The NC and other return streams flow in a northerly direction into Salmon Falls Creek and Snake River, respectively. The Snake River flows to the west.

Published by Burleigh Dodds Science Publishing Limited, 2024.

3.2 Impact of intensive dairy production on regional nutrient balances and losses of reactive nitrogen

Dairy production in the Magic Valley region of Idaho is characterized by high animal density at the regional (~2 AU ha^{-1}) as well as the farm level, where typical densities are closer to 4-20 AU ha^{-1}. Due to the import of nutrients via feed, on-farm surpluses of approximately 11 kg P cow^{-1} year^{-1} and 174 kg N cow^{-1} year^{-1} (including unaccounted for N) have been reported (Hristov et al., 2006; Spears et al., 2003). Unaccounted for N in these budgets are associated with losses of volatile N in housing and manure storage, which are ~50% of N excreted in some cases, resulting in N deficits for on-farm forage production (Leytem et al., 2018). Modeling of representative farms in the region has indicated that 70% of farms cannot meet crop N needs with manure only and need to import fertilizer N, while 80% of dairies produced more P on-farm than can be used by growing crops (Dell et al., 2022). These farm gate P surpluses have led to a buildup of soil P on many producer fields as transportation of manure longer distances is cost prohibitive. The Snake River, which receives runoff from agricultural fields in the region, has many segments of the river which have been identified as impaired with respect to P, resulting in total maximum daily loads (TMDLs) being set at 0.075 mg L^{-1} for total P (TP; IDEQ, 2010). In response, the Idaho State Department of Agriculture (ISDA) has required dairies in the state to develop a nutrient management plan to regulate, in particular, the amount of P being land-applied and evaluate the risk of potential P losses to both surface and groundwater via use of a P Site Index (Leytem et al., 2017).

Leytem et al. (2021) investigated the potential to recycle nutrients generated by dairy cattle in the region within the current agricultural system to better balance nutrients produced with regional crop demands, potentially ameliorating negative environmental impacts. They reported that manure N and P accounted for 45% and 55% of total regional inputs, respectively, with the balance consisting of fertilizer and biological N fixation (15%). There was a regional N surplus of 24 172 MT N year^{-1}, when accounting for all N inputs, compared to crop removal, and accounting for losses of both manure and fertilizer N as NH$_3$ (Fig. 4a). Overall, N losses as NH$_3$ were estimated to be approximately 44 000 MT N year^{-1}, representing 47% of N removed by crops each year. There was a regional P surplus of 8913 MT P year^{-1} (Fig. 4b). However, when assessing P application per ha of cropland reported as receiving manure, the average surplus was 146 kg P ha^{-1} year^{-1}. For both N and P, there was a positive relationship between nutrient surpluses and dairy cattle populations ($r^2 > 0.9$). If manure N could be distributed across the regional cropland, then manure N could supply 44% of crop demand. However, it is important to keep in mind that 100% of the manure N is not available the year it is applied

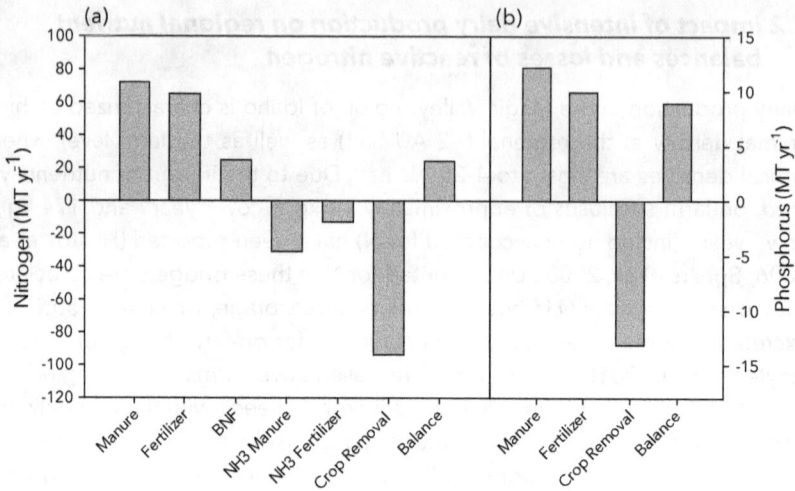

Figure 4 Regional nitrogen (a) and phosphorus (b) balances within the Magic Valley of South Central Idaho, an intensive dairy production region. Balances include N and P applied with manure and fertilizer and removed through crop uptake. Biological nitrogen fixation (BNF) was included as a source of N and losses of N as, ammonia (NH_3), from manure and fertilizer were also included in the balance.

and can vary substantially due to manure type, soils and climate. Average N mineralization rates for this region for solid manure (which represents roughly 67% of total manure) range between 15% and 30% (Leytem, unpublished data) each year. However, over time this N may mineralize and become available for plant uptake or losses to the environment. If N lost as NH_3 could be captured and re-used, then manure N could supply 77% of crop uptake. If manure P were distributed across the regional cropland, it would account for 92% of crop uptake and could replace synthetic fertilizer on most of the acreage. Utilizing the available road system, manure would need to be transported approximately 12.9 km to efficiently cycle it through the regional cropping system. However, due to fields that already have excess P and are not suitable for additional P applications, this transportation distance is likely greater. At present, the average maximum manure transport distance is 6-7 km from the dairy, therefore cost-effective methods for redistribution of manure P from areas of surplus to areas of deficit would need to be developed.

Losses of reactive N in this region have the potential to negatively impact the environment. Regionally, 44 000 MT N year^{-1} is lost as NH_3-N from agricultural production and there is potential for redeposition of N in areas where it could have negative impacts on N sensitive ecosystems. The Magic Valley is surrounded by native rangeland and is upwind of Yellowstone and Grand Teton National Parks, both of which are sensitive to N inputs. Estimating transport and deposition of this NH_3-N has been a new area of investigation in the region.

Utilizing a regional NH_3 monitoring network, estimates of N fluxes indicated a deposition rate of 10 to 40 kg N ha^{-1} year^{-1} in areas impacted by agriculture and dairy production (Leytem et al., unpublished) while the surrounding natural environment may be sensitive to inputs as low as 2 kg ha^{-1} year^{-1}. These additions of N can lead to water quality issues and ecosystem impacts in downwind regions. With a regional N surplus of 24 172 MT N year^{-1}, water quality is also a concern. The USGS evaluated nitrate trends in groundwater over the periods 2000 to 2020 and found increasing trends in regions dominated by intensive agriculture/dairy production, although a majority of wells tested have NO_3 + NO_2 concentrations less than 10 mg L^{-1}, which is below the EPA threshold for drinking water (Skinner, 2022). Lentz et al. (2018) reported that NO_3-N in shallow groundwater in the region increased 1.4-fold from the late 1960s to early 2000s. Even though most well NO_3 concentrations are below the current U.S. EPA threshold, there is concern about increasing trends over time and what that might mean for groundwater quality in the future.

In the Magic Valley region, given the large annual inputs of N, there is potential of significant N_2O losses which can have negative impacts on climate change. To evaluate these potential losses, plot scale research has been conducted within the region to determine the effects of manure/fertilizer application, crop rotation and climate on N_2O losses. In general, fluxes of N_2O-N were greatest when both soil moisture and temperature were high (0.35 m^3 m^{-3} and 25°C), with a few large emission pulses accounting for the majority of growing season N_2O losses (Dungan et al., 2017, 2021; Leytem et al., 2019). In general, the addition of stacked solid manure had greater N_2O emissions than fertilizers or composted manures (Dungan et al., 2017). This enhanced loss of N_2O carried over beyond the year of application with higher losses from treatments receiving manure application 2 years previously. Growing season N_2O emissions also increased significantly with increasing manure application rate (Leytem et al., 2019). The amount of N_2O-N lost as a percentage of the total N applied across multiple studies ranged from <0.01% to 0.41% (Table 1). Regionally, losses of N_2O from application of manure and fertilizer would contribute approximately 1179 MT N_2O-N year^{-1} (assuming 1% loss) to the atmosphere contributing to climate change.

Despite the higher N_2O emissions from manure application, soil organic carbon (SOC) was found to have increased in the soil profile (0-91 cm), while fertilizer treatments had a slight net loss of SOC (Bierer et al., 2021). This increase in SOC can help offset GHG emissions; however, SOC levels begin to decrease once manure applications are terminated, thus the offset will certainly decrease with time since manure was last applied. In addition to providing a potential GHG offset, studies have shown that manure can improve soil health by slowing down or reversing declining organic matter levels. The effect of manure application on soil health was evaluated in the top 30 cm of soil using 12 different biological

Table 1 Emission factors for nitrous oxide losses (N_2O) within the magic valley of southern Idaho

Reference	Treatment	%N_2O-N of TN applied
Dungan et al., 2017	Urea[a]	0.21
	SuperU[b]	0.09
	Compost[c] (34 Mg ha^{-1} annually)	0.09
	Fall manure[d] (52 Mg ha^{-1} annually)	0.09
	Spring manure (52 Mg ha^{-1} annually)	0.12
Leytem et al., 2019	Low manure application rate (18 Mg ha^{-1} annually)	0.13
	Low manure application rate (36 Mg ha^{-1} biennially)	0.15
	High manure application rate (52 Mg ha^{-1} annually)	0.18
	Urea	0.24
Dungan et al., 2021	Urea	<0.01
	Super U	<0.01
	Compost (34 Mg ha^{-1} annually)	0.06
	Fall manure (52 Mg ha^{-1} annually)	0.13
	Spring manure (52 Mg ha^{-1} annually)	0.41

Emission factors are presented as the loss of N_2O-N as a percentage of the total amount of N applied and represent the average (or cumulative) losses over multiple years.
[a] Urea fertilizer only.
[b] SuperU [(stabilized granular urea with urease {N-[n-butyl]-thiophosphoric triamide) and nitrification (dicyandiamide) inhibitors}].
[c] Composted dairy manure scraped from open lots and composted in windrows.
[d] Solid manure scraped from open lot dairies applied in the fall (after harvest) or in the spring (prior to planting) with target application rates of 18, 36 and 52 Mg ha^{-1} annually (dry weight basis).

and chemical indicators (Dungan et al., 2022). These metrics are commonly used to quantify organic matter pools and biological nutrient cycling via enzyme activities and N transformation rates. Compared to synthetic fertilizer, application of manure had a significant positive effect on most indicators, which increased with increasing manure application rate. While these results suggest that manured soil is healthier than soil receiving synthetic fertilizer, the caveat is that applying high rates of manure for many years has resulted in the gradual increase of N and P in the top 30 cm compared to synthetic fertilizer. As previously discussed, this buildup of N and P can pose a threat to water quality.

3.3 Impact of intensive agricultural production on determinants of antibiotic resistance in the environment

While land application of manure can improve soil fertility and health, there are growing concerns over the impact of this practice on the development and dissemination of antibiotic resistance in the environment (Fig. 5). An

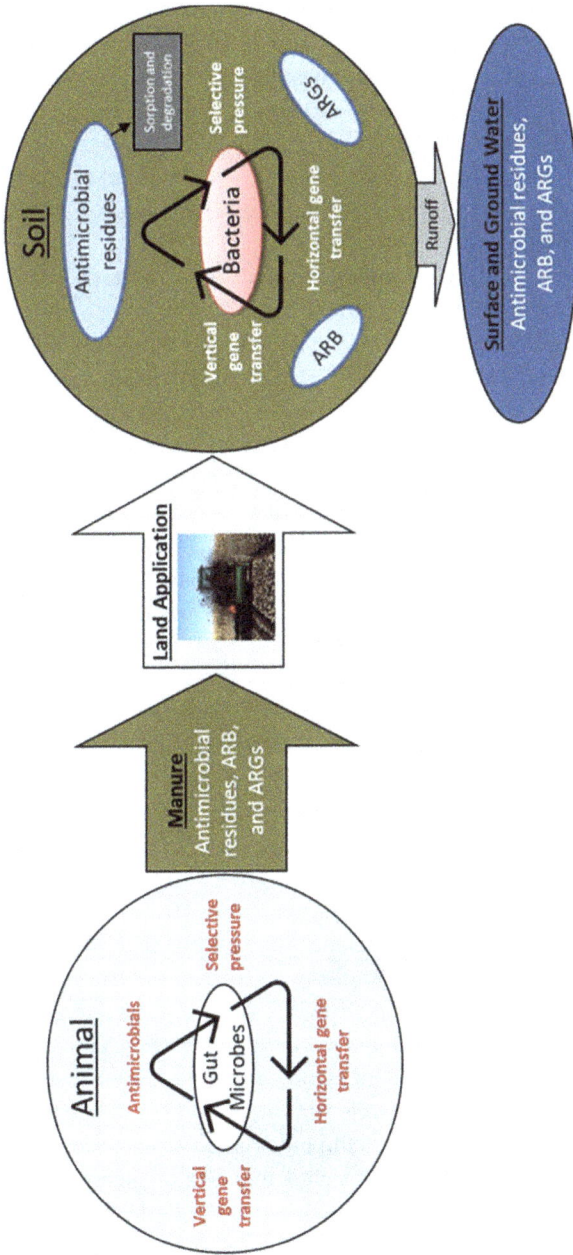

Figure 5 A conceptual diagram showing the potential fate of antimicrobial residues and mechanisms by which antibiotic resistance can be acquired by bacteria within the animal gut and in soil following the land application animal manure. Animal manure application is a major pathway by which antimicrobials, antibiotic resistance genes and bacteria are introduced into the soil environment. Figure adapted from Chee-Sanford et al. (2009).

investigation into the effect of solid dairy manure application on the abundance of ARGs during crop production was performed at plot scale (McKinney et al., 2018). Some ARGs detected at higher relative abundances (normalized to the 16S rRNA gene) in soil with manure application were those encoding resistance to sulfonamide and tetracycline antibiotics. In general, it was found that (1) manure application significantly increased ARG abundance in the top 30 cm of soil compared to plots receiving synthetic fertilizer only, (2) gene abundance increased with increasing manure application rate and (3) subsequent annual applications of manure did not increase the gene abundance above that of the first application.

Dairy wastewater is also commonly applied to agricultural soils; thus, it represents another mechanism by which antibiotic resistance determinants can be disseminated in the environment. To improve knowledge on this topic, a small-scale plot study was established to determine the effect of straight or diluted dairy wastewater on the abundance of ARGs (Dungan et al., 2018). The relative abundance of most ARGs increased dramatically after wastewater irrigation, compared to irrigation of control plots with canal water, and high gene levels were maintained throughout the 6-month study period. The results from this study suggest that increased ARG abundance was by addition of intracellular and extracellular genes present in the wastewater, not by enrichment of ARB. However, it was not known if the presence of antibiotic residues and other xenobiotic compounds may have affected gene selection and persistence in this study. What was clear from these results was that wastewater irrigation dramatically increased ARG abundance in soils receiving straight or diluted wastewater. The increase of the ARG reservoir is a potential cause for concern, as it could facilitate acquired resistance in bacteria that are pathogenic to humans and food-producing animals.

Moving beyond plot-scale studies, research was conducted to determine the abundance of several ARGs in agricultural and non-agricultural soils under various land use practices in the region, including cropland, forestland, inactive cropland, pastureland, rangeland, recreational and residential lands (Fig. 6, Dungan et al., 2019). Soil samples were obtained from 96 sites within seven counties in South Central Idaho. ARGs were detected in many of the soils (15 to 58 out of 96 samples), with a sulfonamide resistance gene being detected the most frequently (60% of samples). All the genes were detected more frequently in the cropland soils (46 sites) and at statistically greater relative abundances than in soils from the other land use categories. When the cropland gene data were separated by sites that had received dairy manure, dairy wastewater and/or biosolids (27 sites), it was revealed that the genes were found at statistically greater abundances (7- to 22-fold higher on average) than in soils that were not treated. This study provided evidence that agricultural production, especially use of manure and biosolids, is contributing to the

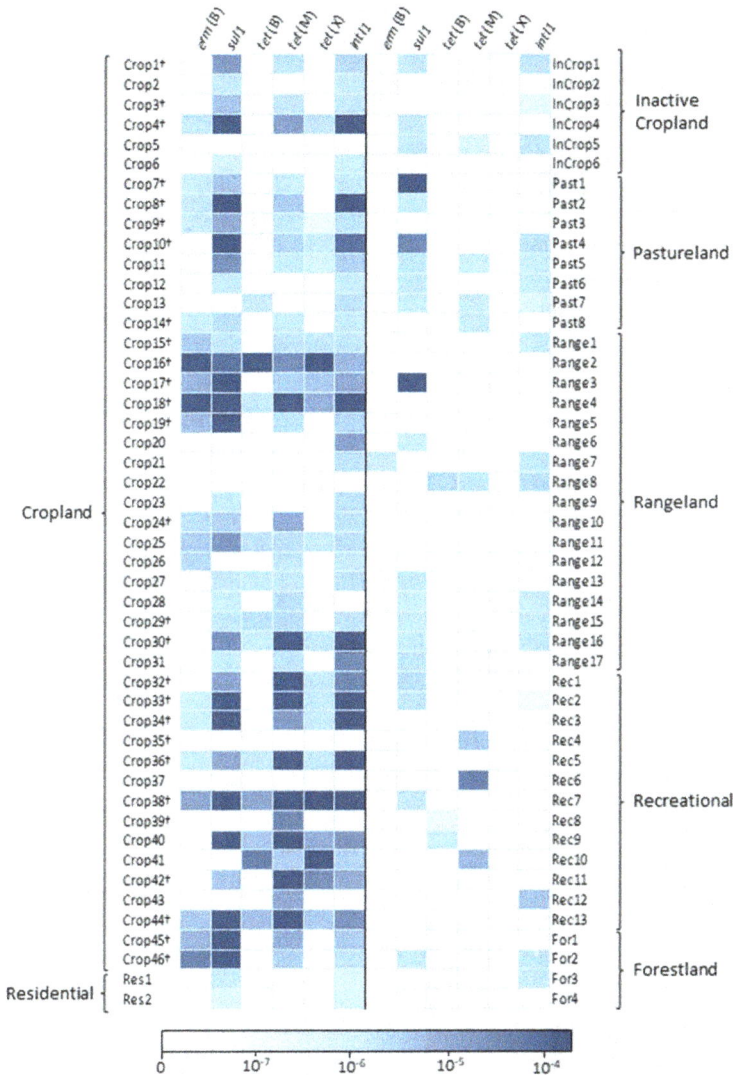

Figure 6 A heat map illustrating the relative abundance of selected antibiotic resistance genes in soils by land use category. The tetracycline resistance genes cover three known mechanisms of resistance including efflux [*tet*(B)], ribosomal protection [*tet*(M)] and enzymatic [*tet*(X)]; *sul1* represents a main determinant of sulfonamide resistance in Gram-negative bacteria; *erm*(B) confers resistance to macrolide-lincosamide-streptogramin B (MLSB) antibiotics and *int1* is a class 1 integron-integrase gene and is linked to antibiotic resistance genes. The color scale at the bottom indicates the gene abundance from no detection (white) to ≥10^{-4} gene copies/16S rRNA gene copies (dark blue). †Indicates cropland soils that have a history of receiving dairy manure, dairy wastewater and/or biosolids.

expansion of antibiotic resistance-related determinants in regional cropland soils.

Land application of manures and biosolids containing antibiotic residues can pose a risk for transport of these residues to surface water via runoff (Joy et al., 2013; Soni et al., 2015; Bartelt-Hunt et al., 2009; Dolliver and Gupta, 2008; Kay et al., 2004). Antibiotics associated with clinical uses can also reach the environment from municipal wastewater treatment plant effluent discharges (Batt et al., 2006; Michael et al., 2013). Regardless of source, antibiotic contamination is recognized as being widespread in water resources. The spatial and temporal occurrence of 21 antibiotics in eight irrigation return flows (IRFs) within the TFCC irrigation tract was investigated (Dungan et al., 2017). Seven antibiotics of veterinary and human origins were detected at frequencies ranging from 3.1% to 62.5%. Monensin, which is exclusively used in cattle and poultry production, was the most frequently detected antibiotic and it was present in all IRFs (Fig. 7). This result suggests that monensin, and other veterinary antibiotics, are entering return flows in runoff from fields that had received livestock manure or wastewater. Some antibiotics were also detected at the site of irrigation water diversion from the Snake River. Therefore, even cropped soils that are not treated with manure, wastewater or biosolids are still receiving low-level antibiotics during irrigation events. The occurrence of antibiotics at low concentrations may pose a health risk to humans and animals due to the proliferation of antibiotic resistance in the environment (Kemper, 2008).

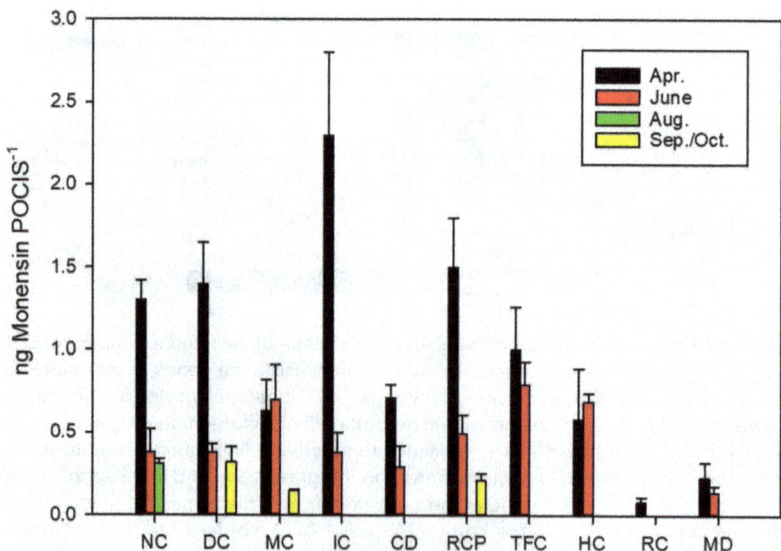

Figure 7 Mass of monensin extracted from polar organic compound integrative samplers (POCIS) that were deployed at the sampling sites within the Upper Snake Rock watershed in South Central Idaho. Columns represent means ± SD (*n* = 3).

Surface waters can be a dominant route by which ARGs and ARB are disseminated in the regional watershed. A study was conducted to determine the abundance of selected ARGs in the IRFs that transport irrigation water from agricultural fields back to the Snake River (Dungan and Bjorneberg, 2020). The ARGs were recovered at all IRF sampling sites with detections ranging from 55 to 81 out of 81 water sampling events. When compared to the average annual relative gene abundances in the canal water samples, they were found to be at statistically greater levels. The fact that most IRFs contained higher levels than found in the canal water indicates that IRFs can be a source of ARGs that ultimately discharge into the Snake River.

The antimicrobial susceptibility among *E. coli* and enterococcal isolates that were obtained from the regional watershed were determined (Dungan and Bjorneberg, 2021). *E. coli* and enterococci are important indicators for understanding the impact of fecal pollution on water resources but reports on antimicrobial resistance among these organisms in IRFs are lacking. Environmental isolates can be a potentially important source of AMR and return flows may be one way resistance genes are transported out of agroecosystems. For *E. coli*, 75% of the isolates were pan-susceptible, while 13% of isolates were resistant to tetracycline and fewer numbers being resistant to 13 other antimicrobials. For the enterococcal species, only 9% of isolates were pan-susceptible and the single highest resistance was to lincomycin (75%). Furthermore, 17 *E. coli* and 13 enterococcal isolates were found to be multidrug resistant to up to seven different drug classes. These results indicate that the IRFs are polluted with material of fecal origin, which would not be surprising given that this is a mixed-use watershed and livestock manures are commonly applied to cropland soils. However, some of the *E. coli* and enterococcal isolates could be from naturalized extraintestinal populations that were released from sediments and soils. Regardless of fecal bacteria source, a wide variety of resistance patterns were found among many of the isolates, suggesting the potential for horizontal transfer of ARGs in the aquatic environment.

4 Conclusion

The valuable nutrients and C found in many agricultural byproducts have the potential to be recycled as bio-based fertilizers within the greater agricultural system, thereby enhancing the circularity and sustainability of this sector. While the potential for beneficial use of these products is great, there are potential risks associated with their use that must be carefully considered. Balancing nutrients in areas with intensive livestock production, avoiding health risks associated with bioaccumulation and biomagnification of heavy metals as well as reducing exposure to pathogens, antibiotics and other CEC are imperative for these products to be used safely and sustainably. Careful risk assessment and management/treatment may overcome most barriers for reuse of these valuable products enabling greater use throughout the agricultural sector.

5 Future trends in research

To further improve the recycling and reuse of agricultural byproducts as bio-based fertilizer in food production systems, further research is needed to address several areas of concern. Improved nutrient use efficiency and reduction of losses, in particular N and P, to the environment are greatly needed. Technologies that allow capture and stabilization of N and P from livestock manure over a range of farm sizes are needed to more effectively recycle these nutrients back to cropland that is deficient in these nutrients. This also includes methods to condense nutrients to reduce the cost of transporting them from regions of surplus to those of deficit. Improved models for estimating crop N uptake to better match N additions are necessary to enhance overall N use efficiencies. Improved models are also needed to estimate N and P losses to the environment to better predict the effects of management strategies on air and water quality as well as losses of N_2O that may impact climate change.

In livestock systems, alternatives to the use of metals (Cu and Zn) and antibiotics in diets as growth promoters are necessary to reduce their potential negative impacts when manure is used as a fertilizer in the landscape. Effective treatment technologies are needed to remove metals, kill pathogens, remove or inactivate antibiotics, and other CECs from agricultural byproducts that may be used as bio-based fertilizers are needed to ensure that these products can be safely and sustainably used within the food production system. Improved risk assessments to establish critical levels of these potential contaminants are also needed to provide guidance for reuse of byproducts as bio-based fertilizers.

6 References

Abascal, E., Gómez-Coma, L., Ortiz, I. and Ortiz, A. (2022). Global diagnosis of nitrate pollution in groundwater and review of removal technologies, *Sci. Total Environ.* 810, 152233. doi: 10.1016/j.scitotenv.2021.152233.

Abu-Ashour, J. and Lee, H. (2000). Transport of bacteria on sloping soil surfaces by runoff, *Environ. Toxicol.* 15(2), 149–153.

Abu-Ashour, J., Joy, D. M., Lee, H., Whiteley, H. R. and Zelin, S. (1994). Transport of microorganisms through soil, *Water Air Soil Pollut.* 75(1–2), 141–158.

Adegoke, A. A., Awolusi, O. O. and Stenström, T. A. (2016). Organic fertilizers: public health intricacies. In: Larramendy, M. L. and Soloneski, S. (Eds), *Organic Fertilizers-From Basic Concepts to Applied Outcomes*. IntechOpen, Croatia, pp. 343–374.

Alegbeleye, O. O., Singleton, I. and Sant'Ana, A. S. (2018). Sources and contamination routes of microbial pathogens to fresh produce during field cultivation: a review, *Food Microbiol.* 73, 177–208.

Alekshun, M. N. and Levy, S. B. (2007). Molecular mechanisms of antibacterial multidrug resistance, *Cell* 128(6), 1037–1050.

Amarakoon, I. D., Surac, S., Zvomuya, F., Cessna, A. J., Larney, F. J. and McAllister, T. A. (2016). Dissipation of antimicrobials in a seasonally frozen soil after beef cattle manure application, *J. Environ. Qual.* 45(5), 1644–1651.

Bahe, A., Classen, J. and Williams, B. (2006). Antibiotic resistance and the environment: Current science and policy considerations. In Rice, J. M. Caldwell, D. F. and Humenik, F. J. (Eds). *Animal Agriculture and the Environment, National Center for Manure & Animal Waste Management White Papers.* American Society of Agricultural and Biological Engineers, St Joseph, pp. 89–108.

Bartelt-Hunt, S. L., Snow, D. D., Damon, T., Shockley, J. and Hoagland, K. (2009). The occurrence of illicit and therapeutic pharmaceuticals in wastewater effluent and surface waters in Nebraska, *Environ. Pollut.* 157(3), 786–791.

Bartelt-Hunt, S. L., Snow, D. D., Damon-Powell, T. and Misebach, D. (2011). Occurrence of steroid hormones and antibiotics in shallow groundwater impacted by livestock waste control facilities, *J. Contam. Hydrol.* 123(3-4), 94–103.

Bateman, A., der Horst, D., Boardman, D., Kansal, A., and Carliell-Marquet, C.(2011). Closing the phosphorus loop in England: the spatio-temporal balance of phosphorus capture from manure versus crop demand for fertilizer, *Res. Conserv. Recyl* 55, 1146–1153.

Batt, A. L., Snow, D. D. and Aga, D. S. (2006). Occurrence of sulfonamide antimicrobials in private water wells in Washington County, Idaho, USA, *Chemosphere* 64(11), 1963–1971.

Baxter-Potter, W. R. and Gilliland, M. W. (1988). Bacterial pollution in runoff from agricultural lands, *J. Environ. Qual.* 17(1), 27–34.

Beneragama, N., Iwasaki, M., Lateef, S. A., Yamashiro, T., Ihara, I. and Umetsu, K. (2013). The survival of multidrug-resistant bacteria in thermophilic and mesophilic anaerobic co-digestion of dairy manure and waste milk, *Anim. Sci. J.* 84(5), 426–433.

Berry, E. D., Woodbury, B. L., Nienaber, J. A., Eigenberg, R. A., Thurston, J. A. and Wells, J. E. (2007). Incidence and persistence of zoonotic bacterial and protozoan pathogens in a beef cattle feedlot runoff control vegetative treatment system, *J. Environ. Qual.* 36(6), 1873–1882.

Bhatnagar, A. and Sillanpää, M. (2011). A review of emerging adsorbents for nitrate removal from water, *Chem. Eng. J.* 168(2), 493–504.

Bierer, A. M., Leytem, A. B., Dungan, R. S., Moore, A. D. and Bjorneberg, D. L. (2021). Soil organic carbon dynamics in semi-arid irrigated cropping systems, *Agronomy* 11(3), 484. doi: 10.3390/agronomy11030484.

BijaySingh, and Craswell, E. (2021). Fertilizers and nitrate pollution of surface and ground water: an increasingly pervasi`ve global problem, *SN Appl. Sci.* 3(4), 518. doi: 10.10 07/s42452-021-04521-8.

Binh, C. T. T., Heuer, H., Kaupenjohann, M. and Smalla, K. (2008). Piggery manure used for soil fertilization is a reservoir for transferable antibiotic resistance plasmids, *FEMS Microbiol. Ecol.* 66(1), 25–37.

Bloem, E., Albihn, A., Elving, J., Hermann, L., Lehmann, L., Sarvi, M., Schaaf, T., Schick, J., Turtola, E. and Ylivainio, K. (2017). Contamination of organic nutrient sources with potentially toxic elements, antibiotics and pathogen microorganisms in relation to P fertilizer potential and treatment options for the production of sustainable fertilizers: a review, *Sci. Total Environ.* 607-608, 225-242.

Bodirsky, B. L., Popp, A., Weindl, I., Dietrich, J. P., Rolinski, S., Scheiffele, L., Schmitz, C. and Lotze-Campen, H. (2012). N$_2$O emission from the global agricultural nitrogen cycle: current state and future scenarios, *Biogeosciences* 9(10), 4169-4197.

Boudh, S. and Singh, J. S. (2019). Pesticide contamination: environmental problems and remediation strategies. In: Bharagave, R. N. and Chowdhary, P. (Eds). *Emerging and Eco-friendly Approaches for Waste Management.* Springer, Singapore, pp. 245-269.

Boxall, A. B. A., Kolpin, D. W., Sorensen, B. H. and Tolls, J. (2003). Are veterinary medicines causing environmental risks?, *Environ. Sci. Technol.* 37(15), 286A-294A.

Boxall, A. B. A., Fogg, L. A., Blackwell, P. A., Kay, P., Pemberton, E. J. and Croxford, A. (2004). Veterinary medicines in the environment, *Rev. Environ. Contam. Toxicol.* 180, 1-91.

Cajamarca, S. M. N., Martins, D., da Silva, J., Fontenelle, M., Guedes, Í., de Figueiredo, C. and Pacheco Lima, C. (2019). Heterogeneity in the chemical composition of biofertilizers, potential agronomic use, and heavy metal contents of different agro-industrial wastes, *Sustainability* 11(7), 1995. doi: 10.3390/su11071995. Available at: www.mdpi.

Callejón, R. M., Rodríguez-Naranjo, M. I., Ubeda, C., Hornedo-Ortega, R., Garcia-Parrilla, M. C. and Troncoso, A. M. (2015). Reported foodborne outbreaks due to fresh produce in the United States and European Union: trends and causes, *Foodb. Pathog. Dis.* 12(1), 32-38.

Campagnolo, E. R., Hohnson, K. R., Karpati, A., Rubin, C. S., Kolpin, D. W., Meyer, M. T., Esteban, J. E., Currier, R. W., Smith, K., Thu, K. M. and McGeehin, M. (2002). Antimicrobial residues in animal water and water resources proximal to large-scale swine and poultry feeding operations, *Sci. Total Environ.* 299(1-3), 89-95.

Center for Disease Control (2020). Foodborne germs and illnesses. Available at: https://www.cdc.gov/foodsafety/foodborne-germs.html (accessed October 17, 2022).

Chang, R. X., Michel Jr., F. C., Gan, J. J., Wang, Q., Wang, Z. Z. and Li, Y. M. (2017). Effect of single and combined herbicides in compost on growth of sensitive crops: green bean, cucumber, and tomato, *Compost Sci. Util.* 25 (sup1), S23-S30.

Chantziaras, I., Boyen, F., Callens, B. and Dewulf, J. (2014). Correlation between veterinary antimicrobial use and antimicrobial resistance in food-producing animals: a report on seven countries, *J. Antimicrob. Chemother.* 69(3), 827-834.

Chaturvedi, P., Shukla, P., Giri, B. S., Chowdhary, P., Chandra, R., Gupta, P. and Pandey, A. (2021). Prevalence and hazardous impact of pharmaceutical and personal care products and antibiotics in environment: a review on emerging contaminants, *Environ. Res.* 194, 110664.

Chee-Sanford, J. C., Mackie, R. I., Koike, S., Krapac, I. G., Lin, Y.-F., Yannarell, A. C., Maxwell, S. and Aminov, R. I. (2009). Fate and transport of antibiotic residues and antibiotic resistance genes following land application of manure waste, *J. Environ. Qual.* 38, 1086-1108. doi: 10.2134/jeq2008.0128.

Christian, T., Schneider, R. J., Färber, H. A., Skutlarek, D., Meyer, M. T. and Goldbach, H. E. (2003). Determination of antibiotic residues in manure, soil, and surface waters, *Acta Hydrochim. Hydrobiol.* 31(1), 36-44.

Coyne, M. S., Gilfillen, R. A., Rhodes, R. W. and Belvins, R. L. (1995). Soil and fecal coliform trapping by grass filter strips during simulated rains, *J. Soil Water Conserv.* 50, 405-408.

Cytryn, E. (2013). The soil resistome: the anthropogenic, the native, and the unknown, *Soil Biol. Biochem.* 63, 18-23.

Davidson, E. A., David, M. B., Galloway, J. N., Goodale, C. L., Haeuber, R., Harrison, J. A., Howarth, R. W., Jaynes, D. B., Lowrance, R. R., Nolan, B. T., Peel, J. L., Pinder, R. W., Porter, E., Snyder, C. S., Townsend, A. R. and Ward, M. H. (2012). Excess nitrogen in the U.S. environment: trends, risks, and solutions. Issues in Ecology, report #15. Ecological Society of America, Washington, DC.

Dell, C. J., Baker, J. M., Spiegal, S., Porter, S. A., Leytem, A. B., Flynn, K. C., Rotz, C. A., Bjorneberg, D. L., Bryant, R. B., Hagevoort, G. R., Williamson, J. C., Slaughter, A. and Kleinman, P. J. A. (2022). Challenges and opportunities for manureshed management across U.S. dairy systems: case studies from four regions, *J. Environ. Qual.* 51(4), 521-539.

Diehl, D. L. and La Para, T. M. (2010). Effect of temperature on the fate of genes encoding tetracycline resistance and the integrase of class 1 integrons within anaerobic and aerobic digesters treating municipal wastewater solids, *Environ. Sci. Technol.* 44(23), 9128-9133.

Dolliver, H. and Gupta, S. (2008). Antibiotic losses in leaching and surface run-off from manure-amended agricultural land, *J. Environ. Qual.* 37(3), 1227-1237.

Doran, J. W. and Linn, D. M. (1979). Bacteriological quality of runoff water from pastureland, *Appl. Environ. Microbiol.* 37(5), 985-991.

Dungan, R. S. (2010). Board-Invited Review: fate and transport of bioaerosols associated with livestock operations and manures, *J. Anim. Sci.* 88(11), 3693-3706.

Dungan, R. S. and Bjorneberg, D. L. (2020). Antibiotic resistance genes, class 1 integrons, and IncP-1/IncQ-1 plasmids in irrigation return flows, *Environ. Pol.L.* 257, 113568, 1-8.

Dungan, R. S. and Bjorneberg, D. L. (2021). Antimicrobial resistance in Escherichia coli and enterococcal isolates from irrigation return flows in a high-desert watershed, *Front. Microbiol.* 12, 660697.

Dungan, R. S., Leytem, A. B., Tarkalson, D. D., Ippolito, J. A. and Bjorneberg, D. L. (2017a). Greenhouse gas emissions from an irrigated cropping system as influenced by nitrogen source and timing, *Soil Sci. Soc. Am. J.* 81(3), 537-545.

Dungan, R. S., Snow, D. D. and Bjorneberg, D. L. (2017b). Occurrence of antibiotics in an agricultural watershed in south-central Idaho, *J. Environ. Qual.* 46(6), 1455-1461.

Dungan, R. S., McKinney, C. W. and Leytem, A. B. (2018). Tracking antibiotic resistance genes in soil irrigated with dairy wastewater, *Sci. Total Environ.* 635, 1477-1483.

Dungan, R. S., Strausbaugh, C. A. and Leytem, A. B. (2019). Survey of selected antibiotic resistance genes in agricultural and non-agricultural soils in south-central Idaho, *FEMS Microbiol. Ecol.* 95(6). doi: 10.1093/femsec/fiz071.

Dungan, R. S., Leytem, A. B. and Tarkalson, D. D. (2021). Greenhouse gas emissions from an irrigated cropping rotation with dairy manure utilization in a semiarid climate, *Agron. J.* 113(2), 1222-1237.

Dungan, R. S., McKinney, C. W., Acosta-Martinez, V. and Leytem, A. B. (2022). Response of soil health indicators to long-term dairy manure in a semiarid irrigated cropping system, *Soil Sci. Soc. Am. J.* 86(6), 1597-1610. doi: 10.1002/saj2.20462.

Elmund, G. K., Morrison, S. M., Grant, D. W. and Nevins, S. M. Sr. (1971). Role of excreted chlorotetracycline in modifying the decomposition process in feedlot waste, *Bull. Environ. Contam. Toxicol.* 6(2), 129-132.

Elser, J. J., Andersen, T., Baron, J. S., Bergström, A. K., Jansson, M., Kyle, M., Nydick, K. R., Steger, L. and Hessen, D. O. (2009). Shifts in lake N:P stoichiometry and nutrient limitation driven by atmospheric nitrogen deposition, *Science* 326(5954), 835–837.

Evans, M. R. and Owens, J. D. (1972). Factors affecting the concentration of faecal bacteria in land-drainage water, *J. Gen. Microbiol.* 71(3), 477–485.

Fajardo, J. J., Bauder, J. W. and Cash, S. D. (2001). Managing nitrate and bacteria in runoff from livestock confinement areas with vegetative filter strips, *J. Soil Water Conserv.* 56, 185–191.

FAO (2006). *Plant Nutrition for Food Security: A Guide for Integrated Nutrient Management.* FAO Fertilizer and Plant Nutrition Bulletin 16. Food and Agriculture Organization of the United Nations, Rome.

Fine, D. D., Breidenbach, G. P., Price, T. L. and Hutchins, S. R. (2003). Quantitation of estrogens in ground water and swine lagoon samples using solid-phase extraction, pentafluorobenzyl/trimethylsilyl derivatizations and gas chromatography-negative ion chemical ionization tandem mass spectrometry, *J. Chromatogr. A* 1017(1–2), 167–185.

Gidden, M. J., Riahi, K., Smith, S. J., Fujimori, S., Luderer, G., Kriegler, E., van Vuuren, D. P., van den Berg, M., Feng, L., Klein, D., Calvin, K., Doelman, J. C., Frank, S., Fricko, O., Harmsen, M., Hasegawa, T., Havlik, P., Hilaire, J., Hoesly, R., Horing, J., Popp, A., Stehfest, E. and Takahashi, K. (2019). Global emissions pathways under different socioeconomic scenarios for use in CMIP6: a dataset of harmonized emissions trajectories through the end of the century, *Geosci. Model Dev.* 12(4), 1443–1475.

Golovko, O., Ahrens, L., Schelin, J., Sörengård, M., Bergstrand, K. J., Asp, H., Hultberg, M. and Wiberg, K. (2022). Organic micropollutants, heavy metals and pathogens in anaerobic digestate based on food waste, *J. Environ. Manage.* 313, 114997.

Gong, Q., Chen, P., Shi, R., Gao, Y., Zheng, S. A., Xu, Y., Shao, C. and Zheng, X. (2019). Health assessment of trace metal concentrations in organic fertilizer in northern China, *Int. J. Environ. Res. Public Health* 16(6), 1031.

Goyal, D., Yadav, A., Prasad, M., Singh, T.B., Shrivastav, P., Ali, A., Dantu, P. K. and Mishra, S. (2020). Effect of heavy metals on plant growth: an overview. In: Naeem, M., Ansari, A. and Gill, S. (Eds). *Contaminants in Agriculture.* Springer Nature, Switzerland.

Guan, J., Wasty, A., Grenier, C. and Chan, M. (2007). Influence of temperature on survival and conjugative transfer or multiple antibiotic-resistant plasmids in chicken manure and compost microcosms, *Poult. Sci.* 86(4), 610–613.

Guo, R. X. and Chen, J. Q. (2015). Application of alga-activated sludge combined systems (AASCS) as a novel treatment to remove cephalosporins, *Chem. Eng. J.* 260, 550–556.

Gurtler, J. B., Doyle, M. P., Erickson, M. C., Jiang, X., Millner, P. and Sharma, M. (2018). Composting to inactivate foodborne pathogens for crop soil application: a review, *J. Food Prot.* 81(11), 1821–1837.

Halling-Sørensen, B., Nielsen, S. N., Lanzky, P. F., Ingerslev, F., Holten Lützhøft, H. C. and Jørgensen, S. E. (1998). Occurrence, fate and effects of pharmaceutical substances in the environment - a review, *Chemosphere* 36(2), 357–393.

Halling-Sørensen, B., Sengeløv, G. and Tjørnelund, J. (2002). Toxicity of tetracyclines and tetracycline degradation products to environmentally relevant bacteria, including selected tetracycline-resistant bacteria, *Arch. Environ. Contam. Toxicol.* 42(3), 263–271.

Hamscher, G., Sczesny, S., Höper, H. and Nau, H. (2002). Determination of persistent tetracycline residues in soil fertilized with liquid manure by high-performance liquid chromatography with electrospray ionization tandem mass spectrometry, *Anal. Chem.* 74(7), 1509–1518.

Hanserud, O. S., Brod, E., Øgaard, A. F., Müller, D. B. and Brattebø, H. (2016). A multi-regional soil phosphorus balance for exploring secondary fertilizer potential: the case of Norway, *Nutr. Cycl. Agroecosyst.* 104(3), 307–320.

Hao, C., Lissemore, L., Nguyen, B., Kleywegt, S., Yang, P. and Solomon, K. (2006). Determination of pharmaceuticals in environmental waters by liquid chromatography/electrospray ionization/tandem mass spectrometry, *Anal. Bioanal. Chem.* 384(2), 505–513.

Heinonen-Tanski, H., Mohaibes, M., Karinen, P. and Koivunen, J. (2006). Methods to reduce pathogen microorganisms in manure, *Livest. Sci.* 102(3), 248–255.

Herrero, M. and Thornton, P. K. (2013). Livestock and global change: emerging issues for sustainable food systems, *Proc. Natl Acad. Sci. U. S. A.* 110(52), 20878–20881.

Heuer, H. and Smalla, K. (2007). Manure and sulfadiazine synergistically increased bacterial antibiotic resistance in soil over at least two months, *Environ. Microbiol.* 9(3), 657–666.

Holly, M. A., Kleinman, P. J., Bryant, R. B., Bjorneberg, D. L., Rotz, C. A., Baker, J. M., Boggess, M. V., Brauer, D. K., Chintala, R., Feyereisen, G. W., Gamble, J. D., Leytem, A. B., Reed, K. F., Vadas, P. A. and Waldrip, H. M. (2018). Short communication: identifying challenges and opportunities for improved nutrient management through the USDA's Dairy Agroecosystem Working Group, *J. Dairy Sci.* 101(7), 6632–6641.

Holm-Nielsen, J. B., Al Seadi, T. and Oleskowicz-Popiel, P. (2009). The future of anaerobic digestion and biogas utilization, *Bioresour. Technol.* 100(22), 5478–5484.

Honorio, J. F., Veit, M. T. and Tavares, C. R. G. (2019). Alternative adsorbents applied to the removal of natural hormones from pig farming effluents and characterization of the biofertilizer, *Environ. Sci. Poll. Res. Int.* 26(28), 28429–28435.

Hristov, A. N., Hazen, W. and Ellsworth, J. W. (2006). Efficiency of use of imported nitrogen, phosphorus, and potassium and potential for reducing phosphorus imports on Idaho dairy farms, *J. Dairy Sci.* 89(9), 3702–3712.

Hutchins, S. R., White, M. V., Hudson, F. M. and Fine, D. D. (2007). Analysis of lagoon samples from different concentrated animal feeding operations for estrogens and estrogen conjugates, *Environ. Sci. Technol.* 41(3), 738–744.

Hutchison, M. L., Walters, L. D., Avery, S. M., Munro, F. and Moore, A. (2005). Analyses of livestock production, waste storage, and pathogen levels and prevalences in farm manures, *Appl. Environ. Microbiol.* 71(3), 1231–1236.

Idaho Department of Environmental Quality. Issued permits and water quality certification. Available at: https://www.deq.idaho.gov/permits/issued-permits-and-water-quality -certifications/ (accessed 17 October 2022).

Jamieson, R. C., Gordon, R. J., Sharples, K. E., Stratton, G. W. and Madani, A. (2002). Movement and persistence of fecal bacteria in agricultural soils and subsurface drainage waters: a review, *Can. Biosyst. Eng.* 44, 1.1–1.9.

Jenkins, M. B., Endale, D. M., Schomberg, H. H. and Sharpe, R. R. (2006). Fecal bacteria and sex hormones in soil and runoff from cropped watersheds amended with poultry litter, *Sci. Total Environ.* 358(1–3), 164–177.

Jiang, Y., Xie, S. H., Dennehy, C., Lawlor, P. G., Hu, Z. H., Wu, G. X., Zhan, X. M. and Gardiner, G. E. (2020). Inactivation of pathogens in anaerobic digestion systems for converting biowastes to bioenergy: a review, *Renew. Sustain. Energy Rev.* 120, 109654.

Joy, S. R., Bartelt-Hunt, S. L., Snow, D. D., Gilley, J. E., Woodbury, B. L., Parker, D. B., Marx, D. B. and Li, X. (2013). Fate and transport of antimicrobials and antimicrobial resistance genes in soil and runoff following land application of swine manure slurry, *Environ. Sci. Technol.* 47(21), 12081–12088.

Kay, P., Blackwell, P. A. and Boxall, A. B. (2004). Fate of veterinary antibiotics in a macroporous tile drained clay soil, *Environ. Toxicol. Chem.* 23(5), 1136–1144.

Kemper, N. (2008). Veterinary antibiotics in the aquatic and terrestrial environment, *Ecol. Indic.* 8(1), 1–13.

Klein, M., Brown, L., Tucker, R. W., Ashbolt, N. J., Stuetz, R. M. and Roser, D. J. (2010). Diversity and abundance of zoonotic pathogens and indicators in manures of feedlot cattle in Australia, *Appl. Environ. Microbiol.* 76(20), 6947–6950.

Koelsch, R., Lorimor, J. and Mankin, K. (2006). 'Vegetative treatment systems for open lot runoff: Review of literature. In Rice, J. M., Caldwell, D. F. and Humenik, F. J. (Eds). *Animal Agriculture and the Environment, National Center for Manure & Animal Waste Management White Papers*. American Society of Agricultural and Biological Engineers, St Joseph, MI, pp. 576–608.

Kraus, H., Weber, A., Appel, M., Enders, B., Isenberg, H. D., Schiefer, H. G., Slenczka, W., von Graevenitz, A. and Zahner, H. (2003). *Zoonoses: Infectious Diseases Transmissible from Animals to Humans*. ASM Press, Washington, DC.

Lentz, R. D., Carter, D. L. and Haye, S. V. (2018). Changes in groundwater quality and agriculture in forty years on the Twin Falls irrigation tract in southern Idaho, *J. Soil Water Conserv.* 73(2), 107–119.

Létourneau, V., Duchaine, C., Côté, C., Letellier, A., Topp, E. and Massé, D. (2010). Presence of zoonotic pathogens in physico-chemically characterized manure from hog finishing houses using different production systems, *Bioresour. Technol.* 101(11), 4048–4055.

Levy, S. B. (1992). *The Antibiotic Paradox. How Miracle Drugs Are Destroying the Miracle*. Plenum Press, New York.

Leytem, A., Bjorneberg, D. and Tarkalson, D. (2017). The phosphorus site index: A systematic approach to assess the risk of nonpoint source pollution of Idaho waters by agricultural phosphorus. Available at: https://agri.idaho.gov/main/wp-content/uploads/2018/12/Phosphorus-Site-Index-reference-2017-revised.pdf (accessed 14 October 2022).

Leytem, A. B., Bjorneberg, D. L., Rotz, C. A., Moraes, L. E., Kebreab, E. and Dungan, R. S. (2018). Ammonia emissions from dairy lagoons in the western U.S., *Trans. ASABE* 61(3), 1001–1015.

Leytem, A. B., Moore, A. D. and Dungan, R. S. (2019). Greenhouse gas emissions from an irrigated crop rotation utilizing dairy manure, *Soil Sci. Soc. Am. J.* 83(1), 137–152.

Leytem, A. B., Williams, P., Zuidema, S., Martinez, A., Chong, Y. L., Vincent, A., Vincent, A., Cronan, D., Kliskey, A., Wulfhorst, J. D., Alessa, L. and Bjorneberg, D. (2021). Cycling phosphorus and nitrogen through cropping systems in an intensive dairy production region, *Agronomy* 11(5), 1005.

Li, H., Sun, Z., Qiu, Y., Yu, X., Han, X. and Ma, Y. (2018). Integrating bioavailability and soil aging in the derivation of DDT criteria for agricultural soils using crop species sensitivity distributions, *Ecotoxicol. Environ. Saf.* 165, 527–532.

Liu, M., Huang, X., Song, Y., Tang, J., Cao, J., Zhang, X., Zhang, Q., Wang, S., Xu, T., Kang, L., Cai, X., Zhang, H., Yang, F., Wang, H., Yu, J. Z., Lau, A. K. H., He, L., Huang, X., Duan, L., Ding, A., Xue, L., Gao, J., Liu, B. and Zhu, T. (2019). Ammonia emission control in China would mitigate haze pollution and nitrogen deposition, but worsen acid rain, *Proc. Natl Acad. Sci. U. S. A.* 116(16), 7760-7765.

Liu, Q., Wang, J., Bai, Z., Ma, L. and Oenema, O. (2017). Global animal production and nitrogen and phosphorus flows, *Soil Res.* 55(6), 451-462.

Liu, L., Xu, W., Zhong, B., Lu, X., Zhong, B., Guo, Y., Lu, X., Zhao, Y., He, W., Wang, S., Zhang, X., Liu, X. and Vitousek, P. (2022). Exploring global changes in agricultural ammonia emissions and their contribution to nitrogen deposition since 1980, *PNAS* 119(14), e2121998119. doi: 10.1073/pnas.2121998119.

Lopes, C., Herva, M., Franco-Uría, A. and Roca, E. (2011). Inventory of heavy metal content in organic waste applied as fertilizer in agriculture: evaluating the risk of transfer into the food chain, *Environ. Sci. Pollut. Res. Int.* 18(6), 918-939.

Lund, B., Jensen, V. F., Have, P. and Ahring, B. (1996). Inactivation of virus during anaerobic digestion of manure in laboratory scale biogas reactors, *Anton. Leeuw.* 69(1), 25-31.

Luo, L., Ma, Y., Zhang, S., Wei, D. and Zhu, Y. G. (2009). An inventory of trace element inputs to agricultural soils in China, *J. Environ. Manage.* 90(8), 2524-2530.

Ma, R., Li, K., Guo, Y., Zhang, B., Zhao, X., Linder, S., Guan, C., Chen, G., Gan, Y. and Meng, J. (2021). Mitigation potential of global ammonia emissions and related health impacts in the trade network, *Nat. Commun.* 12(1), 6308.

Manyi-Loh, C. E., Mamphweli, S. N., Meyer, E. L., Makaka, G., Simon, M. and Okoh, A. I. (2016). An overview of the control of bacterial pathogens in cattle manure, *Int. J. Environ. Res. Public Health* 13(9), International, J. Environ. Res, *Pub Health* 13, 843.

Martens, W. and Böhm, R. (2009). Overview of the ability of different treatment methods for liquid and solid manure to inactivate pathogens, *Bioresour. Technol.* 100(22), 5374-5378.

Matthews, K. R. (2006). *Microbiology of Fresh Produce.* ASM Press, Washington, DC.

Matthiessen, P., Allen, Y., Bamber, S., Craft, J., Hurst, M., Hutchinson, T., Feist, S., Katsiadaki, I., Kirby, M., Robinson, C., Scott, S., Thain, J. and Thomas, K. (2002). The impact of oestrogenic and androgenic contamination on marine organisms in the United Kingdom-summary of the EDMAR programme. Endocrine disruption in the marine environment, *Mar. Environ. Res.* 54(3-5), 645-649.

Mawdsley, J. L., Bardgett, R. D., Merry, R. J., Pain, B. F. and Theodorou, M. K. (1995). Pathogens in livestock waste, their potential movement through soil and environmental pollution, *Appl. Soil Ecol.* 2(1), 1-15.

McKinney, C. W., Dungan, R. S., Moore, A. and Leytem, A. B. (2018). Occurrence and abundance of antibiotic resistance genes in agricultural soil receiving dairy manure, *FEMS Microbiol. Ecol.* 94(3), 1-10.

Michael, I., Rizzo, L., McArdell, C. S., Manaia, C. M., Merlin, C., Schwartz, T., Dagot, C. and Fatta-Kassinos, D. (2013). Urban wastewater treatment plants as hotspots for the release of antibiotics in the environment: a review, *Water Res.* 47(3), 957-995.

Naghdi, M., Taheran, M., Brar, S. K., Kermanshahi-Pour, A., Verma, M. and Surampalli, R. Y. (2018). Removal of pharmaceutical compounds in water and wastewater using fungal oxidoreductase enzymes, *Environ. Pollut.* 234, 190-213.

Negreanu, Y., Pasternak, Z., Jurkevitch, E. and Cytryn, E. (2012). Impact of treated wastewater irrigation on antibiotic resistance in agricultural soils, *Environ. Sci. Technol.* 46(9), 4800–4808.

Nekvapil, F., Ganea, I. V., Ciorîță, A., Hirian, R., Ogresta, L., Glamuzina, B., Roba, C. and Cintă Pinzaru, S. (2021). Wasted biomaterials from crustaceans as a compliant natural product regarding microbiological, antibacterial properties and heavy metal content for reuse in blue bioeconomy: A preliminary study, *Materials (Basel)* 14(16), 4558.

Nesme, T., Senthilkumar, K., Mollier, A. and Pellerin, S. (2015). Effects of crop and livestock segregation on phosphorus resource use: A systematic, regional analysis, *Eur. J. Agron.* 71, 88–95.

Nguyen, T. T., Rosello, C., B´elanger, R.and Ratti, C. (2020). Fate of residual pesticides in fruit and vegetable waste (FVW) processing, *Foods* 9(10), 1468.

Nightingale, K. K., Schukken, Y. H., Nightingale, C. R., Fortes, E. D., Ho, A. J., Her, Z., Grohn, Y. T., McDonough, P. L. and Wiedmann, M. (2004). Ecology and transmission of Listeria monocytogenes infecting ruminants and in the farm environment, *Appl. Environ. Microbiol.* 70(8), 4458–4467.

Nunes, N., Ragonezi, C., Gouveia, C. S. S. and Pinheiro de Carvalho, M. Â. A. (2021). Review of sewage sludge as a soil amendment in relation to current international guidelines: A heavy metal perspective, *Sustainability* 13(4), 2317.

O'Connor, J., Hoang, S. A., Bradney, L., Dutta, S., Xiong, X., Tsang, D. C. W., Ramadass, K., Vinu, A., Kirkham, M. B. and Bolan, N. S. (2021). A review on the valorisation of food waste as a nutrient source and soil amendment, *Environ. Pollut.* 272, 115985.

O'Connor, J., Mickan, B. S., Siddique, K. H. M., Rinklebe, J., Kirkham, M. B. and Bolan, N. S. (2022). Physical, chemical, and microbial contaminants in food waste management for soil application: a review, *Environ. Pollut.* 300, 118860.

Parchomenko, A. and Borsky, S. (2018). Identifying phosphorus hot spots: A spatial analysis of the phosphorus balance as a result of manure application, *J. Environ. Manage.* 214, 137–148.

Patel, A. B., Shaikh, S., Jain, K. R., Desai, C. and Madamwar, D. (2020). Polycyclic aromatic hydrocarbons: sources, toxicity, and remediation approaches, *Front. Microbiol.* 11, 562813.

Pepper, I. L., Brooks, J. P. and Gerba, C. P. (2006). Pathogens in biosolids, *Adv. Agron.* 90, 1–41.

Polechónska, L., Klink, A., Dambiec, M. and Rudecki, A. (2018). Evaluation of *Ceratophyllum demersum* as the accumulative bioindicator for trace metals, *Ecol. Indic.* 93, 274–281.

Prather, M. J., Hsu, J., DeLuca, N. M., Jackman, C. H., Oman, L. D., Douglass, A. R., Fleming, E. L., Strahan, S. E., Steenrod, S. D., Søvde, O. A., Isaksen, I. S., Froidevaux, L. and Funke, B. (2015). Measuring and modeling the lifetime of nitrous oxide including its variability, *J. Geophys. Res. Atmos.* 120(11), 5693-5705.

Prathumchai, N., Polprasert, C. and Englande Jr., A. J. (2018). Phosphorus distribution and loss in the livestock sector - the case of Thailand, *Resour. Conserv. Recycl.* 136, 257-266.

Pruden, A., Pei, R. T., Storteboom, H. and Carlson, K. H. (2006). Antibiotic resistance genes as emerging contaminants: studies in northern Colorado, *Environ. Sci. Technol.* 40, 7445-7450.

Raghothama, K. G. (2005). Phosphorus and plant nutrition: an overview. In: Sims, J. T. and Sharpley, A. N. (Eds). *Phosphorus: Agriculture and the Environment, Agronomy Monograph NO. 46 ASA, CSSA, SSSA*. Madison Book Company, Madison, pp. 355–378.

Ramos, M. C., Quinton, J. N. and Tyrrel, S. F. (2006). Effects of cattle manure on erosion rates and runoff water pollution by faecal coliforms, *J. Environ. Manage.* 78(1), 97–101.

Rathi, B. S., Kumar, P. S. and Show, P. L. (2021). A review on effective removal of emerging contaminants from aquatic systems: current trends and scope for further research, *J. Hazard. Mater.* 409, 124413.

Ratnam, S., March, S. B., Ahmed, R., Bezanson, G. S. and Kasatiya, S. (1988). Characterization of Escherichia coli serotype O157:H7, *J. Clin. Microbiol.* 26(10), 2006–2012.

Resende, J. A., Silva, V. L., Rocha de Oliveira, T. L., de Oliveira Fortunato, S., da Costa Carneiro, J., Otenio, M. H. and Diniz, C. G. (2014). Prevalence and persistence of potentially pathogenic and antibiotic resistant bacteria during anaerobic digestion treatment of cattle manure, *Bioresour. Technol.* 153, 284–291.

Sandegren, L. (2014). Selection of antibiotic resistance at very low antibiotic concentrations, *Ups. J. Med. Sci.* 119(2), 103–107.

Sardar, K., Ali, S., Hameed, S., Afzal, S., Fatima, S., Shakoor, M. B., Bharwana, S. A. and Tauqeer, H. M. (2013). Heavy metals contamination and what are the impacts on living organisms, *Greener J Environ. Manag. Public Saf* 2(4), 172–179.

Sarmah, A. K., Meyer, M. T. and Boxall, A. B. A. (2006). A global perspective on the use, sales, exposure pathways, occurrence, fate and effects of veterinary antibiotics (VAs) in the environment, *Chemosphere* 65(5), 725–759.

Sethy, S. K. and Ghosh, S. (2013). Effect of heavy metals on germination of seeds, *J. Nat. Sci. Biol. Med.* 4(2), 272–275.

Shi, T., Ma, J., Wu, X., Ju, T., Lin, X., Zhang, Y., Li, X., Gong, Y., Hou, H., Zhao, L. and Wu, F. (2018). Inventories of heavy metal inputs and outputs to and from agricultural soils: a review, *Ecotoxicol. Environ. Saf.* 164, 118–124.

Skinner, K. D. (2022)-5060. *Trends in Groundwater Levels, and Orthophosphate and Nitrate Concentrations in the Middle Snake River Region, South-Central Idaho*. Scientific Investigations Report. United State Geologic Survey, Reston, VA.

Smit, A. L., van Middelkoop, J. C., van Dijk, W. and van Reuler, H. (2015). A substance flow analysis of phosphorus in the food production, processing and consumption system of the Netherlands, *Nutr. Cycl. Agroecosyst.* 103(1), 1–13.

Sobsey, M. D., Khatib, L. A., Hill, V. R., Alocilja, E., Pillai, S. (2006). 'Pathogen in animal wastes and the impacts of waste management practices on their survival, transport and fate'. In Rice, J. M., Caldwell, D. F. and Humenik, F. J. (Eds). Animal Agriculture and the Environment, National Center for Manure & Animal Waste Management White Papers. American Society of Agricultural and Biological Engineers, St. Joseph, MI, pp. 609–666.

Song, W., Huang, M., Rumbeiha, W. and Li, H. (2007). Determination of amprolium, carbadox, monensin, and Tylosin in surface water by liquid chromatography/tandem mass spectrometry, *Rapid Commun. Mass SPectrom.* 21(12), 1944–1950.

Soni, B., Bartelt-Hunt, S. L., Snow, D. D., Gilley, J. E., Woodbury, B. L., Marx, D. B. and Li, X. (2015). Narrow grass hedges reduce Tylosin and associated antimicro-bial resistance genes in agricultural runoff, *J. Environ. Qual.* 44(3), 895–902.

Spears, R. A., Kohn, R. A. and Young, A. J. (2003). Whole-farm nitrogen balance on western dairy farms, *J. Dairy Sci.* 86(12), 4178–4186.

Sutton, M. A., Mason, K. E., Sheppard, L. J., Sverdrup, H., Haeuber, R. and Hicks, W. K. (2014). Nitrogen deposition, critical loads and biodiversity. Springer, Dordretch, Netherlands.

Svanbäck, A., McCrackin, M. L., Swaney, D. P., Linefur, H., Gustafsson, B. G., Howarth, R. W. and Humborg, C. (2019). Reducing agricultural nutrient surpluses in a large catchment-Links to livestock density, *Sci. Total Environ.* 648, 1549–1559.

Tan, L. V. and Tran, T. (2021). Heavy metal accumulation in soil and water in pilot scale rice field treated with sewage sludge, *ChemEngineering* 5(4), 77.

Tester, C. F. (1990). Organic amendment effects on physical and chemical properties of a sandy soil, *Soil Sci. Soc. Am. J.* 54(3), 827–831.

Thakali, A. and MacRae, J. D. (2021). A review of chemical and microbial contamination in food: what are the threats to a circular food system?, *Environ. Res.* 194, 110635.

Tian, H., Xu, R., Canadell, J. G., Thompson, R. L., Winiwarter, W., Suntharalingam, P., Davidson, E. A., Ciais, P., Jackson, R. B., Janssens-Maenhout, G., Prather, M. J., Regnier, P., Pan, N., Pan, S., Peters, G. P., Shi, H., Tubiello, F. N., Zaehle, S., Zhou, F., Arneth, A., Battaglia, G., Berthet, S., Bopp, L., Bouwman, A. F., Buitenhuis, E. T., Chang, J., Chipperfield, M. P., Dangal, S. R. S., Dlugokencky, E., Elkins, J. W., Eyre, B. D., Fu, B., Hall, B., Ito, A., Joos, F., Krummel, P. B., Landolfi, A., Laruelle, G. G., Lauerwald, R., Li, W., Lienert, S., Maavara, T., MacLeod, M., Millet, D. B., Olin, S., Patra, P. K., Prinn, R. G., Raymond, P. A., Ruiz, D. J., van der Werf, G. R., Vuichard, N., Wang, J., Weiss, R. F., Wells, K. C., Wilson, C., Yang, J. and Yao, Y. (2020). A comprehensive quantification of global nitrous oxide sources and sinks, *Nature* 586(7828), 248–256.

Tian, L., Jinjin, C., Ji, R., Ma, Y. and Yu, X. (2022). 'Microplastics in agricultural soils: sources, effects, and their fate', *Curr. Opin, Environ. Sci. Health* 25, 100311.

Tiquia, S. M., Tam, N. F. Y. and Hodgkiss, I. J. (1998). Salmonella elimination during composting of spent pig litter, *Bioresour. Technol.* 63(2), 193–196.

Tran, N. H., Reinhard, M. and Gin, K. Y. H. (2018). Occurrence and fate of emerging contaminants in municipal wastewater treatment plants from different geographical regions-a review. *Water Res.* 133, 182–207.

Tyrrel, S. F. and Quinton, J. N. (2003). Overland flow transport of pathogens from agricultural land receiving faecal wastes, *J. Appl. Microbiol.* 94 (Suppl.), 87S–93S.

Ugulu, I., Akhter, P., Khan, Z. I., Akhtar, M. and Ahmad, K. (2021). Trace metal accumulation in pepper (*Capsicum annuum* L.) grown using organic fertilizers and health risk assessment from consumption, *Food Res. Int.* 140, 109992.

Unc, A. and Goss, M. J. (2004). Transport of bacteria from manure and protection of water resources, *Appl. Soil Ecol.* 25(1), 1–18.

UNEP and WHRC (2007). *Reactive Nitrogen in the Environment: Too Much or Too Little of a Good Thing.* United Nations Environment Programme, Paris.

Venglovsky, J., Sasakova, N. and Placha, I. (2009). Pathogens and antibiotic residues in animal manures and hygienic and ecological risks related to subsequent land application, *Bioresour. Technol.* 100(22), 5386–5391.

Wang, R., Goll, D., Balkanski, Y., Hauglustaine, D., Boucher, O., Ciais, P., Janssens, I., Penuelas, J., Guenet, B., Sardans, J., Bopp, L., Vuichard, N., Zhou, F., Li, B., Piao, S., Peng, S., Huang, Y. and Tao, S. (2017). Global forest carbon uptake due to nitrogen and phosphorus deposition from 1850 to 2100, *Glob. Change Biol.* 23(11), 4854–4872.

Waseem, H., Williams, M. R., Stedtfeld, R. D. and Hashsham, S. A. (2017). Antimicrobial resistance in the environment, *Water Environ. Res.* 89(10), 921-941.

Weber, R., Herold, C., Hollert, H., Kamphues, J., Blepp, M. and Ballschmiter, K. (2018). Reviewing the relevance of dioxin and PCB sources for food from animal origin and the need for their inventory, control and management, *Environ. Sci. Eur.* 30(1), 42.

Westerman, P. W. and Bicudo, J. R. (2005). Management considerations for organic waste use in agriculture, *Bioresour. Technol.* 96(2), 215-221.

Williams-Nguyen, J., Sallach, J. B., Bartelt-Hunt, S., Boxall, A. B., Durso, L. M., McLain, J. E., Singer, R. S., Snow, D. D. and Zilles, J. L. (2016). Antibiotics and antibiotic resistance in agroecosystems: state of the science, *J. Environ. Qual.* 45(2), 394-406.

Yang, Q., Tian, H., Li, X., Ren, W., Zhang, B., Zhang, X. and Wolf, J. (2016). Spatiotemporal patterns of livestock manure nutrient production in the conterminous United States from 1930 to 2012, *Sci. Total Environ.* 541, 1592-1602.

Youngquist, C. P., Mitchell, S. M. and Cogger, C. G. (2016). Fate of antibiotics and antibiotic resistance during digestion and composting: a review, *J. Environ. Qual.* 45(2), 537-545.

Yu, Y., Zhou, Y., Wang, Z., Torres, O. L., Guo, R. and Chen, J. (2017). Investigation of the removed mechanisms of antibiotic ceftazidime by green algae and subsequent microbic impact assessment, *Sci. Rep.* 7, 1-11.

Zhan, X., Bo, Y., Zhou, F., Liu, X., Paerl, H. W., Shen, J., Wang, R., Li, F., Tao, S., Dong, Y. and Tang, X. (2017). Evidence for the importance of atmospheric nitrogen deposition to eutrophic lake Dianchi, China, *Environ. Sci. Technol.* 51(12), 6699-6708.

Zhan, X., Adalibieke, W., Cui, X., Winiwarter, W., Reis, S., Zhang, L., Bai, Z., Wang, Q., Huang, W. and Zhou, F. (2021). Improved estimates of ammonia emissions from global croplands, *Environ. Sci. Technol.* 55(2), 1329-1338.

Zhang, B., Tian, H., Lu, C., Dangal, S. R. S., Yang, J. and Pan, S. (2017). Global manure nitrogen production and application in cropland during 1860-2014: a 5 arcmin gridded global dataset for Earth system modeling, *Earth Syst. Sci. Data* 9(2), 667-678.

Chapter 4

Biofertilizers: assessing the effects of arbuscular mycorrhizal fungi on soil health

M. J. Salomon, The Waite Research Institute and The School of Agriculture, Food and Wine, The University of Adelaide, Australia; S. F. Bender, Agroscope, Switzerland; T. R. Cavagnaro, The Waite Research Institute and The School of Agriculture, Food and Wine, The University of Adelaide, Australia; and M. G. A. van der Heijden, Agroscope and University of Zurich, Switzerland

1   Introduction
2   Arbuscular mycorrhizal fungi and soil health: addressing the key issues
3   Arbuscular mycorrhizal fungi biofertilizer production
4   Managing arbuscular mycorrhizal fungi for soil health
5   Conclusion
6   Future trends in research
7   Where to look for further information
8   References

1 Introduction

Soils provide a variety of important ecosystem services and are the foundation of global biogeochemical cycles such as carbon, water and plant nutrients. They host an abundance of microorganisms, ranging from the microscopic to the macroscopic level (Adhikari and Hartemink, 2016). Healthy soils have the capacity to provide those ecosystem functions that are appropriate to its surrounding and to do so in a sustainable way. As such, healthy soils are the foundation of most food production systems, ecosystems and urban settlements (Keesstra et al., 2016, Salomon et al., 2020). Many developments of the past and present have led to land loss and land degradation. Anthropogenic activities that heavily impact soils on a global scale include intensive agriculture, deforestation and land disturbances, such as for urban settlements or mining. Those impacts can be complex and, as many systems are interwoven with one another, can result in a cascade of events with broader

http://dx.doi.org/10.19103/AS.2021.0094.17

implications on its surroundings. Estimates suggest that up to 45% of global land areas are degraded, which may undermine the well-being of 1.5 billion people (Gibbs and Salmon, 2015). Major drivers of land degradation are soil erosion, acidification, land clearance, salination or pollution through heavy metals and petrochemicals (Olsson et al., 2019).

The previous decades saw a variety of concepts and names that were used for soil assessment. Among those are soil fertility, soil quality, soil capability and, recently, soil health. Similarly, the objectives and methods of each concept progressed throughout the years. The earliest forms of soil assessment were emphasizing the suitability for crop growth, whereas newer concepts consider the multifunctionality of whole ecosystem services. Soil assessment can be done on soil biology (e.g. microbial communities), chemistry (e.g. pH and nutrients) and physics (e.g. bulk density and texture). Current developments in soil analysis allowed a shift towards the inclusion of soil biology as commonly used indicators, which were previously more focused on soil chemistry and physics. Following, we adhere to the term 'soil health' to describe sustainable and resilient ecosystem services (Bünemann et al., 2018).

The sustainability of many current agricultural practices has been questioned. For example, one consequence of suboptimal soil management is the release of soil carbon into the atmosphere which is about the same magnitude as carbon emissions caused by current deforestation events (Sanderman et al., 2017). The concentration of carbon in the soil is tightly linked with important soil characteristics, including soil aggregation and soil microbial biomass (Wilson et al., 2009b). As a consequence, the loss of soil carbon leads to deteriorated soil health and contributes to elevated atmospheric carbon dioxide (CO_2) concentrations (Rumpel et al., 2020). Intense farming systems put further pressure on soil fertility and can result in nutrient depletion. Such soils lack the capacity of replenishing essential plant nutrients, and crop yields are only sustained through the application of fertilizers (Tan et al., 2005). As a further consequence, soils might see a loss of biodiversity which deprives them of their essential ecosystem functions (van der Heijden et al., 2008).

Soils are home to a wide range of biota that have beneficial effects on plant growth and soil functioning. One of those groups is the mycorrhizal fungi. Mycorrhizal fungi follow a cosmopolitan distribution and are one of the main drivers of soil microbial interactions. Their hyphae can build up to 40 m g^{-1} of soil (Smith et al., 2004) and reach a biomass of 700-900 kg ha^{-1} (Wallander et al., 2001). Mycorrhizal fungi can be broadly categorized into ectomycorrhizas, ericoid mycorrhizas, orchid mycorrhizas and arbuscular mycorrhizas (Smith and Read, 2008). These mycorrhizas differ in their physiology, symbiotic strategies and taxonomic classification. However, they all form mycorrhizal associations with host plants that can be described as a mutualistic symbiosis. Only in particular cases or circumstances are these associations parasitic (Smith and Read, 2008).

Arbuscular mycorrhizas are plant-fungal associations in which the fungus enters the plant root and forms specialized structures, the arbuscules, which are used for nutrient exchange with the host plant (see Fig. 1). Arbuscular mycorrhizal fungi (AMF) are obligate symbionts for 80% of terrestrial plant species and most crop plants. This mutualistic relationship is improving plant nutrition by aiding in the uptake of phosphorus and zinc. In return, the plant delivers photosynthates and lipids to the fungal symbiont (Smith and Read, 2008). Besides improved plant nutrition, AMF play a crucial role in soil health and are key indicators for describing soil quality. AMF have been found to improve many aspects of soil health and counteract the negative impacts caused by inappropriate soil management (Jeffries et al., 2003). At the same time, common agricultural practices led to a diminished abundance of AMF in the soil. Such practices include simplified crop rotations, application of mineral fertilizer or soil disturbances, such as cultivation (Verbruggen et al., 2010).

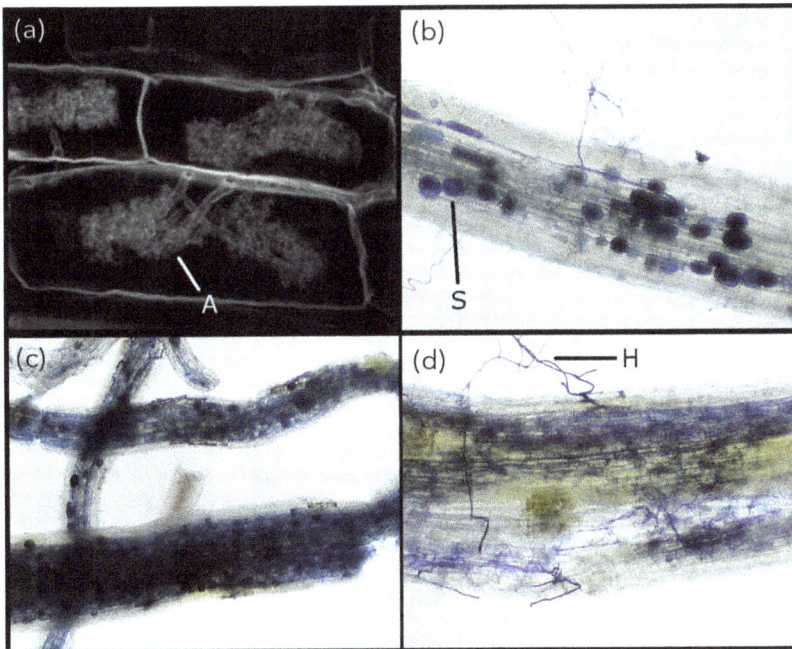

Figure 1 (a) Mycorrhizal arbuscules inside Pisum sativum root cells; (b-d) colonized maize roots with many intraradical spores and extraradical hyphae. A = arbuscule, S = spore, H = hyphae. Figure 1 (a): Photo courtesy of Ryan Geil, published with kind permission from Peterson and Massicotte (2004) and NRC Press, © Canadian Science Publishing or its licensors.

One proposed way to increase soil health in a sustainable way is the application of AMF as a biofertilizer. AMF biofertilizers are designed to bolster natural mycorrhizal communities when those have been impaired or to introduce new mycorrhizal isolates with improved functional traits. The AMF inoculum is embedded in a carrier material which might include further additives, such as organic additives (e.g. humic acids) or other plant growth-promoting microorganisms such as *Trichoderma spp.*, *Bacillus spp.* or other microorganisms. Such biofertilizers have been proposed as an alternative to mineral fertilizers and pesticides and have been known to increase yield resilience by supporting plants against abiotic stress (Berruti et al., 2016). While AMF biofertilizers hold enormous potential to improve sustainability, their integration into broad-scale applications has been challenging. The mass production of AMF propagules is defined by its symbiotic lifecycle in which mycorrhizal fungi require a host plant to propagate. This production method is linked to phytosanitary issues and requires much care to exclude plant pathogens. Although AMF can be cultured axenically, it is currently not economical to do so on a broad scale. Furthermore, most root culture systems require genetically modified hairy roots, which are produced using tumour-inducing (Ti) plasmids (Berruti et al., 2016, Adholeya et al., 2005). Recent studies demonstrated that AMF cannot produce fatty acids themselves and rely on their host plants for providing them. Thus, it is possible that specific growth media might be designed in the future, in which it is possible to grow AMF without host plants (Kameoka et al., 2019). Further issues arise when applying AMF inoculum to field soil, which already contains an established microbial community. Under those circumstances, survival and establishment of the introduced AMF needs to be evaluated (Rodriguez and Sanders, 2015, Bender et al., 2019). Furthermore, AMF show a certain host selectivity and might not perform consistently in a crop rotation (Hoeksema et al., 2010).

Although the production and application of AMF biofertilizer come with certain challenges, an increasing number of companies are attracted by its potential economic value (Vosátka et al., 2012). The global market for microbial inoculants is expected to reach US$3.622 billion by 2024, of which the mycorrhizal fungi are a major sector (Knowledge Sourcing Intelligence LLP, 2019). In most countries, the term 'biofertilizer' is not legally defined and is therefore lacking regulations and minimum standard requirements. This situation led to an unregulated market in which a high percentage of commercial AMF biofertilizers fail to induce mycorrhizal colonization. To this date, a number of scientific studies are available where the majority of tested inoculants showed unsatisfying results (Tarbell and Koske, 2007, Corkidi et al., 2017) (Salomon et al., in press). This situation is undermining the potential of AMF biofertilizer with bigger implications for sustainability in food production and consumer protection. Scientific research might be sabotaged when researchers rely on commercial mycorrhizal inoculants.

AMF biofertilizers hold the potential to significantly reduce our carbon footprint on this earth. The management of AMF can be a sustainable option to reduce agrochemicals and to increase soil health at the same time. As outlined above, the production and application of AMF biofertilizers can be challenging and needs further research before they can be adapted on a broader scale. However, those challenges should not be a reason to undermine the potential of AMF, nor should the interest in AMF applications be belittled as a 'recurring evolution' (Hart et al., 2018). Instead, the management and understanding of the soil microbiome should be treated as a valuable tool for humanity to stay within its planetary boundaries (Rockström et al., 2009). Following, we will address the key issues of how AMF can help to improve soil health and outline promising developments towards the use of AMF biofertilizers (see Table 1).

2 Arbuscular mycorrhizal fungi and soil health: addressing the key issues

2.1 Improved soil structure and stability

Soil structure is defined as the spatial arrangement of soil particles to form a three-dimensional matrix consisting of mineral and organic particles (aggregates) and porous spaces in between. Soil stability describes the disintegration forces necessary to disrupt this matrix. Soil structure is a key indicator of soil health as it influences a variety of important soil characteristics. Well-structured soils facilitate root growth, aeration and water infiltration into deeper soil layers, while providing water retention at the same time. High soil stability prevents soil from wind and water erosion. Soil structure and stability are generally influenced by soil physical, chemical and microbial influences. Within the microbial influences, AMF are one of the main contributors towards soil structure and stability.

The effects of AMF on soil structure are of direct and indirect nature. The direct effects include processes in which fungal mycelium enmeshes soil particles into bigger units, including soil microaggregates (<0.25 mm) and macroaggregates (>0.25 mm) (Miller and Jastrow, 1990). The indirect effects are describing how AMF can influence plants and microbial communities which then influence the soil structure (Tisdall and Oades, 1982). At the macroaggregate level, soil structure through AMF is mainly improved due to the physical force provided by the hyphal entanglement of soil particles (Miller and Jastrow, 2000). As mentioned before, AMF hyphae can build up to 40 m g^{-1} of soil (Smith et al., 2004). At the same time, the hyphal diameter is about ten times smaller than that of fine roots, allowing hyphae to penetrate even micropores (<30 μm) (Smith and Read, 2008). The well-known biochemical compound Glomalin and the group of Glomalin-related soil proteins are a biochemical pathway for soil aggregation. This group of proteins is thought to act like a glue for soil particles (Driver et al.,

Table 1 Impact of AMF on soil health: overview of soil ecosystem functions provided by AMF and their underlying mechanism

Soil ecosystem function	Mechanism	Reference
Uptake of plant nutrients and stimulation of plant growth	Improved uptake of phosphorus, nitrogen and micronutrients	Watts-Williams and Cavagnaro (2012)
	Increased plant biomass	van der Heijden et al. (1998)
	Improved drought resistance	Sanchez-Diaz and Honrubia (1994)
	Protective effects against root diseases	Hol and Cook (2005)
	Induction of systemic pathogen resistance	Pozo and Azcón-Aguilar (2007)
Improved soil structure and stability	Hyphal entanglement of soil particles	Miller and Jastrow (1990)
	Carbon deposition	Rillig and Mummey (2006)
	Changes in surface polarity of soil particles	Rillig (2005)
	Alignment of soil particles	Tisdall (1991)
	Eliminating spatial constraints	Six et al. (2004)
	Indirect effects through changes in plant and microbial communities	Rillig and Mummey (2006)
Alleviation of soil contamination	Immobilization of metals	French (2017)
	Upregulation of plant detoxification genes	Jiang et al. (2016)
	Removal of contaminants from plants through fungal structures	Göhre and Paszkowski (2006)
	Increased plant vigour through improved plant nutrition	Vogel-Mikuš and Regvar (2006)
	Degradation of organic pollutants	Wang et al. (2020)
Carbon sequestration	Significant carbon sink into the soil	Douds et al. (2000)
	Increased plant community productivity	Zhang et al. (2012)
	Improved soil structure	Wilson et al. (2009)
Nutrient retention	Improved nutrient uptake of plants	Smith and Read (2008)
	Immobilization in fungal structures	Watts-Williams and Cavagnaro (2012)
	Improved soil structure	Cavagnaro et al. (2015)

2005). However, there are open questions regarding their quantification and release into the soil and if they are specific to AMF (Rosier et al., 2006). Fungal growth is further linked to a variety of other extracellular organic compounds that have been shown to improve the soil structure, for example, through changes in the surface polarity (Gebbink et al., 2005) or carbon deposition. Mycorrhizal effects on the formation of microaggregates are less researched but hypothesized to work through physical forces on primary soil particles. The turgor pressure of hyphae during their growth could eliminate spatial constraints that would otherwise prevent the formation of microaggregates. Similarly, this physical force could align particles and bind them with organic matter (Rillig and Mummey, 2006).

Indirect effects of AMF on soil structure describe its effect on plants and soil communities, which, in return, can influence soil structure and stability. The influence of plant communities on the soil structure has been repeatedly shown, for example, in the context of agricultural (Munkholm et al., 2013) or natural ecosystems (Pérès et al., 2013). The host-selectivity of AMF thereby promotes certain plant species over others, which ultimately leads to changes in soil structure (van der Heijden et al., 2006). The effects of AMF on single plants are mainly evolving around an increase in the ratio of root to shoot biomass as a result of the mycorrhizal symbiosis. Following is a cascade of events with potential positive outcomes on the soil structure, such as increased rhizodeposition, soil entanglement by fine roots, increased root decomposition and changes to the soil water regime (Piotrowski et al., 2004). AMF can have further effects on the composition and quantity of soil microorganisms, which in return lead to changes to the soil structure (van der Heijden et al., 2008, Rillig and Mummey, 2006).

2.2 Soil contamination

Soils are considered contaminated once they contain substances (e.g. organic pollutants), elements (e.g. toxic metals) or microorganisms (e.g. pathogens) above normal concentrations or above national guideline levels (FAO, 2015). Soil contamination is a diverse topic which makes it difficult to quantify it on a global scale. The extent of soil contamination is nevertheless alarming and considered one of the major threats to soil functioning. Although soil contamination can be of natural origin, anthropogenic activities are considered as the main causes (FAO, 2015).

One of the main issues of soil contamination is the accumulation of toxic metals, for example, through the application of agrochemicals, biosolids, wastewater or mining (Wuana and Okieimen, 2011). The earliest studies about the protective effects of mycorrhizal fungi against toxic metals involved ericoid mycorrhizal fungi (Bradley et al., 1981). Since then it has been repeatedly demonstrated that also AMF can alleviate stress caused by increased metal concentrations in soils (Hildebrandt et al., 2007). It is almost a confusing phenomenon, since some metals are necessary micronutrients for plant life and their uptake is promoted by AMF (Watts-Williams and Cavagnaro, 2012). This uptake of micronutrients and the protective effect against toxic levels demonstrate the complex interaction between soils, the soil microbiome and plants. Studies at contaminated sites revealed the presence of AMF communities (Vogel-Mikuš et al., 2005), albeit their diversity is often reduced compared to non-contaminated sites. However, those remaining AMF species might be better adapted to high concentrations of toxic metals (Del Val et al., 1999, Galli et al., 1994). The underlying mechanisms are either changing the fate of the metals in the soil or change the plant's response towards them. One established mechanism is the immobilization of metals in intra- and extraradical fungal structures through metallothioneins or other chelating agents (Lanfranco, 2007, French, 2017). Interactions between metals and Glomalin-related soil proteins have been reported as well, leading to similar forms of immobilization (Yang et al., 2017). Other proposed mechanisms are related to changes in the plant physiology which are caused by the AM symbiosis. These changes involve the upregulation of plant genes involved in detoxification (Jiang et al., 2016) or the removal of contaminants from plant roots through fungal structures such as arbuscules (Göhre and Paszkowski, 2006). More recent studies also investigated the effects of AMF on organic pollutants in soil. The mechanisms involved are similar to those of potentially toxic metals but also include the degradation of pollutants through enzymes. Again, the overall protective effect is composed of multiple mechanisms and interactions between soil microorganisms and plants (Wang et al., 2020). The effects of many other common pollutants such as pesticides and micro-plastics are still poorly investigated. Especially the effects of mixed pesticides in combination with other soil stress factors on soil health are still poorly understood (Rillig et al., 2019b).

2.3 Carbon sequestration

Soils are the largest terrestrial carbon storage. The amount of carbon that is stored in soils exceeds that of the atmosphere and the global plant biomass combined (Scharlemann et al., 2014). The carbon cycle is one of the most important processes on Earth to which almost every organism is linked in one way or another. Plants are one of the main drivers of this process and as such also their symbionts at the root–soil interface. Plants forming associations with AMF allocate up to 20% of carbon to AMF and plants forming ectomycorrhizas can allocate up to 50% of their assimilates, demonstrating that mycorrhizal associations are substantially involved in the carbon cycle (Soudzilovskaia et al., 2015). The prehistoric events that saw a decline in atmospheric carbon are linked to the emergence of deeply rooted trees around 450 million years ago and angiosperms around 130 million years ago. Mycorrhizal associations have been confirmed in root fossils dating back more than 400 million years ago (Remy et al., 1994) and are, therefore, considered crucial for the evolution of terrestrial plants, indicating that their role in carbon cycling might be of equal importance (Taylor et al., 2009). Soil carbon concentration is an important factor for soil health and is the foundation of almost all other soil components. Having high concentrations of soil carbon is therefore considered valuable for 'healthy soils'. However, due to unsustainable management, soils are increasingly releasing carbon into the atmosphere in the form of CO_2. This development is aggravating climate change when soils could actually be used as carbon storage and bind atmospheric CO_2 into the ground. One initiative predicts that an annual growth rate of 0.4% soil carbon could have significant effects on soil health and contribute to limit global warming at the same time (Kon Kam King et al., 2018).

Studying the impact of AMF on the soil carbon cycle is a difficult undertaking on an ecosystem level. Pot studies under controlled conditions allow to separate the carbon inputs between roots and AMF. However, on an ecosystem level, this becomes increasingly complicated, as an almost unmanageable amount of processes interact with AMF and vice versa. On the level of a single plant, mycorrhizal associations are translocating photosynthates from the plant directly into the soil. Hexose is the preferred metabolite for AMF which is sourced from plants and then used for growth and reproduction. Estimates suggest that between 5% and 20% of plant-derived carbon is translocated to AMF associations (Douds and Shachar-Hill, 2000). The overall carbon sink through AMF mycelium thereby accumulates to 50–900 kg carbon ha^{-1} (Zhu and Michael Miller, 2003). Although this carbon drain might suggest negative impacts on plant growth, most mycorrhizal plants react to mycorrhizal colonization by producing more biomass. AMF provide many advantages such as improved nutrient uptake which then leads to higher photosynthetic rates. Another mechanism on how AMF improve carbon sequestration is their positive effect on soil stability and soil aggregation. Mycorrhizas are critical components of the terrestrial carbon cycle and shape plants and soils alike. Conversely, depriving soils of their mycorrhizal potential leads to implications that go beyond the effects on a single plant (Wilson et al., 2009a).

2.4 Nutrient retention

High concentrations of plant nutrients can be challenging when those are not fully integrated into the ecosystem. Nutrient availability that is exceeding its demand leads to nutrient loss, either through leaching, gaseous emissions or erosion. This situation is mainly observed after the application of fertilizers, for example, in the agricultural context. Nutrients that are mobile within the soil matrix can leach out and make them inaccessible to the root system. Various forms of nitrogen are at risk of being lost through gaseous emissions, such as in the form of nitrous oxide (N_2O). Immobile nutrients such as some forms of phosphorus are more susceptible to soil erosion than leaching. Although phosphorus can also bind to mobile particles and then leach through the soil. Such excess nutrients can have negative impacts on their surroundings. Leached or wind-dispersed nutrients contaminate water ways and groundwater, thereby damaging the aquatic biodiversity. N_2O is a potent greenhouse gas with significant global warming potential. The overall fertilizer efficiency is sobering, especially when considering the global scale of fertilizer application. It is estimated that 30% of nitrogen and 15–30% of phosphorus are lost in the same year of application (Quan et al., 2020, Syers et al., 2008).

One method by which AMF can improve nutrient retention is obviously due to its role in plant nutrition, especially in the case of phosphorus. Without mycorrhizal associations, plant roots are inside a depletion zone once all immobile nutrients (like phosphorus) around the roots are taken up, making them dependent on nutrients moving to roots via mass flow, usually a slow process. Plants with mycorrhizal colonization can escape this depletion zone through the hyphae which can penetrate a bigger soil volume than roots by itself. This way, phosphorus gets bound into fungal and plant biomass which immobilizes it and protects it from erosion and leaching. Furthermore, mycorrhiza show enhanced mineralization of complex-bound or organic forms of phosphorus which would otherwise be unavailable to plant roots and susceptible of leaching. Another essential plant nutrient that is impacted by AMF is nitrogen. To this date, the exact interactions between AMF and the various forms of nitrogen are still to be investigated (Hodge and Fitter, 2010). However, it is known that AMF can take up nitrogen in the form of ammonium (NH_4^+) and amino acids. Evidence suggests that even other organic forms of nitrogen might be involved, especially since AMF are equipped with nitrogen reductase genes. The net balance of AMF and its role in nitrogen uptake on plants differ between studies. Whereas some studies show almost no contribution, other studies come to the opposite result. Although the effect of AMF on the nitrogen uptake of the plant is variable, its impact on reducing nitrogen losses has been proven repeatedly (Cavagnaro et al., 2015).

Another important mechanism on how AMF help in nutrient retention is through effects on soil and plant water relations. There is some evidence that AMF affect plant water transport and can support plant water uptake (Augé et al., 2015, Bowles et al., 2016). Enhanced plant water uptake will consequently reduce the amount of leachate and the amount of nutrients being transported down the soil profile, which can also reduce N_2O emissions (Lazcano et al., 2014). In addition, the above-mentioned effects of AMF on enhanced soil aggregation can lead to significant effects on water retention. Both mechanisms together directly impact the amount of leaching in soils (Cavagnaro et al., 2015). Altogether, the impact of nutrient retention due to AMF is compelling. Nutrient retention of up to 80% has been reported for nitrogen and up to 60% for phosphorus (van der Heijden, 2010, Corkidi et al., 2011). Gaseous emissions of N_2O were reduced between about 30% and 50% after the application of nitrate fertilizer (Bender et al., 2014, 2015).

3 Arbuscular mycorrhizal fungi biofertilizer production

As outlined in the previous sections, mycorrhizal fungi have been linked to increased soil health and increased plant vigour. They are key players in the soil microbiome and at the soil–root interface. These characteristics make them promising alternatives to agrochemicals which can be used in sustainable agriculture and ecosystem restoration. The idea of using AMF as biofertilizers is almost as old as the systematic research of AMF itself. However, most work to transfer the scientific research into efficient applications has been rather futile up to this date. AMF biofertilizers are still considered a niche product and are mostly limited to the hobby market or some horticulture systems (Rouphael et al., 2015). The main reasons for the narrow application range being high costs (Berruti et al., 2016), unreliable product quality (Salomon et al., in press) and questions regarding the establishment under field conditions (Rodriguez and Sanders, 2015). Regardless of those issues, their huge environmental and economic potential is fuelling continuous research and major investments from agrochemical companies (Vosátka et al., 2008).

The earliest systems for the propagation of AMF used host plants in pot cultures and sterilized soil, which were then inoculated with the desired species. Preferred host plants are maize or sorghum, as their fast-growing root system allows substantial sporulation (Ijdo et al., 2011). To this date, it is still the most common method for mass production of mycorrhizal fungi, as it is relatively cheap and can be easily upscaled (Ijdo et al., 2011). This *in vivo* system underwent multiple modifications, such as the closed bag system which would reduce the costs for maintenance and helped to exclude potential phytopathogens (Walker and Vestberg, 1994). Multiple publications describe the use of soil-less substrate, such as sand (Jentschke et al., 1999), which could further help with phytosanitary issues and eliminate the need for soil sterilization. It is even possible to produce on-site inoculum on agricultural waste products, yielding 3600 spores in 100 mL^{-1} of soil (Chaiyasen et al., 2017). The simplicity of *in vivo* systems allows easy propagation and maintenance of mycorrhizal cultures. Major bottlenecks for the use in commercial biofertilizer production are phytosanitary issues and the difficulty to extract spores from the substrate, for example, to be further concentrated or to be used with different carrier materials (see Fig. 2a).

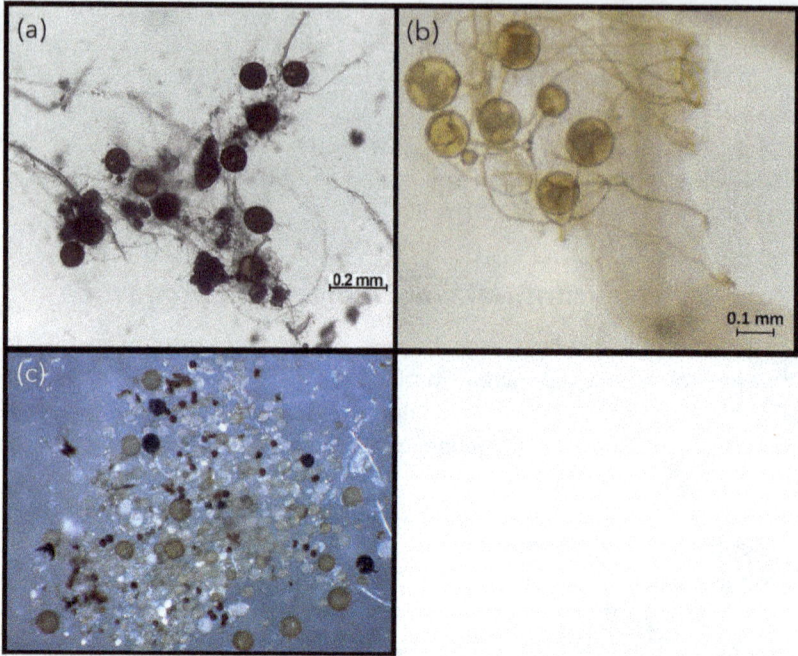

Figure 2 (a) Extracted AMF spores from a commercial product; (b) spores of an in vitro culture of Rhizophagus clarus growing on tomato roots; (c) spores of various sizes and colours as can be found in natural soils. Photo courtesy of Luise Köhl.

The need for axenic AMF systems led to the development of *in vitro* cultures on Ti-transformed hairy roots (Declerck et al., 1996) (see Fig. 2b). The motivation behind this development was not only its potential use for the production of biofertilizers but also providing a research tool for the study of AMF. The original method made it possible to produce up to 10 000 spores of *Glomus versiforme* within 4 months. The spore production could be further increased by using a split-plate design (65 000 spores) (Douds et al., 2000) or whole plants instead of hairy roots (Voets et al., 2005). The latest research even demonstrated the asymbiotic sporulation of *Rhizophagus irregularis* using culture medium with fatty acids (Kameoka et al., 2019). AMF propagules in axenic systems can be easily extracted and are free of plant pathogens by default. They can be added to a variety of carrier materials which makes them applicable for most production systems. However, axenic cultures require trained personal and high material costs which make it too expensive for most biofertilizer applications. Also, not all AMF species can be cultured *in vitro* and continuous subculturing has been shown to lead to changes in functional traits (Kokkoris and Hart, 2019). Some producers question the ability of axenic cultures to establish in the field (e.g. because the AMF did not adapt and grow under real soil conditions), and this is an issue that needs further testing.

Other approaches for the production of AMF inoculum include hydroponic or aeroponic systems. To this date, there have been a variety of adaptations, each with customized designs and slightly different nutrient solutions. One of the earliest works used the nutrient flow technique (NFT) to produce *Glomus mosseae* propagules on maize, yielding up to 50% mycorrhizal root infection (Elmes and Mosse, 1984). This system provided good growth conditions with relatively little root disturbance to protect the fungal structures. But also deep water culture systems with reduced aeration times have been successfully used (Hawkins and George, 1997). Aeroponic systems have certain advantages over hydroponics, such as increased root aeration which supports plant growth and reduces the likelihood of certain diseases. Nutrients and moisture are applied as aerosols around the roots. Common systems convert the nutrient solution into micro-droplets (spray nozzles), mist (atomizers) or ultrasonic fog (piezo elements) (Jarstfer and Sylvia, 1995). With regard to the production of mycorrhizal inoculum, aeroponics put less physical stress onto the root systems than hydroponics. Production rates of up to 50 spores cm^{-1} of colonized roots have been reported (Hung and Sylvia, 1988). Aeroponics and hydroponics allow easy spore extraction through sieving the roots and moist roots can be further processed to sheared-root inoculum using a common food blender. Sheared-root inoculum can only be made from wet roots and then stored for up to 1 month, whereas unprocessed and dried roots retain their inoculum potential for up to 2 years, but are not suitable for the production of sheared-root inoculum (Sylvia and Jarstfer, 1992). Those examples provide proof of the flexibility of AMF production systems and explain how they can be advanced to fulfil specific requirements. Most companies do not disclose their production methods due to issues about intellectual property. However, we can assume that successful companies developed efficient production systems which might be based on the mentioned hydroponic or aeroponic systems. Systems that have been made public involve an airlift bioreactor (Jolicoeur et al., 1999) or a semi-hydroponic set-up (Declerck et al., 2009).

4 Managing arbuscular mycorrhizal fungi for soil health

4.1 Agriculture

The management of AMF in agriculture has been subject to recent discussions which were mainly focused on insufficient correlations between the yield of agricultural crops and their mycorrhizal colonization (Ryan and Graham, 2018). It is obvious that many plants can grow without AMF assuming that sufficient amounts of fertilizers are added. However, it is important to consider the whole spectrum of ecosystem services which are provided by AMF, especially those related to soil health (Chen et al., 2018). Those additional services are often not as visible as crop yields but nevertheless of high economic and ecological importance (Rillig et al., 2019a). Especially in times of extreme weather events, good soil health and functioning mycorrhizal symbiosis can be considered as a 'health insurance' for crops to achieve yield resilience (Rivero et al., 2018). However, such things are often difficult to measure, scientifically and monetarily. Moreover, AMF can help to reduce the reliance on fertilizers and make agriculture more sustainable. Future research should focus on the role that AMF can play for soil ecological engineering, producing the same amount of food with less inputs.

Mycorrhizal colonization of crop plants through indigenous AMF can be increased through changes in agricultural practices. Key variables which have been identified are the use of cover crops and the type of tillage system (Bowles et al., 2017). Those results indicate that mycorrhizal colonization of summer crops increases by 30% when working with minimal soil disturbance and cover crops instead of barrow fallows. Furthermore, the AMF species richness increased by 11%, highlighting the sensitivity of certain species to soil disturbance. Further research focused on the comparison between organic and conventional agriculture, whereas organic systems are usually associated with higher AMF abundance and richness. Again, tillage and cover crops have been considered as major influences, as well as the impact of fertilization (Borriello et al., 2012). Mineral fertilizers are generally associated with reduced AMF abundance and species richness, caused by their high availability and effects on soil chemistry (Oehl et al., 2004). Most organic farms also incorporate more diverse crop rotations and take advantage of the many positive effects of legumes and mixed crops on soil health (Verbruggen et al., 2010).

Research shows that soils can be actively managed to support AMF and increase soil health. Such measures include organic farming, permanent crop cover, crop rotation and the inclusion of temporary pastures in rotations. However, these methods come with limitations. Organic farming is relying on ploughing and tilling as means to control weeds and phytopathogens. Arid farmlands in many parts of the world do not support multiple crops within one season. It is especially those harsh environments that could benefit from the many advantages of AMF. Canola is an important cash crop in organic and conventional farming systems alike. However, being a non-mycorrhizal plant, it has negative impacts on following crops in terms of mycorrhizal colonization (Valetti et al., 2016). Where options to manage indigenous AMF are limited or weakened by non-mycorrhizal crops, soils can be actively inoculated with AMF biofertilizer. This inoculum could be adapted to the local conditions and provide optimized plant and ecosystem functions (Sanders, 2010).

Although AMF provide multiple ecosystem services, they need to provide clear economic benefits to find their way into most broad-scale applications.

Assuming that most farmers do not get direct fiscal support to protect their soils, the costs of biofertilizer need to be weighed against the cost of potential fertilizer-savings or yield and quality gains. Major factors for this cost-benefit analysis are the amount of necessary inoculum and the specific mycorrhizal growth response of the crop. Due to its relatively low seed density, potato is a crop with strong prospects for profitable AMF biofertilization. Hijri (2016) demonstrated in field trials over 4 years an average yield increase of about 4 t ha^{-1}. The profitability threshold was reached in almost 80% of all trials while an average of 71 spores was applied per seed potato at the time of planting. Tawaraya et al. (2012) achieved significant savings in phosphorus fertilization for the Welsh onion (*Allium fistulosum*). Typical for this crop, the seeds were pre-planted under greenhouse conditions where seedlings were either treated with *Glomus etunicatum* 'R10' or left uninoculated. Plants were then transplanted into the field under various fertilization treatments. The yield of the inoculated plants at 300 mg P$_2$O$_5$ kg^{-1} soil was similar to those in 1000 mg P$_2$O$_5$ kg^{-1} soil and not inoculated, whereas the costs of the difference in superphosphate application were double the costs of the inoculum. Not part of this equation is the savings in environmental costs due to potential fertilizer run-off and/or leaching. For future research, it is important that on-farm experiments are being performed under real-world agricultural conditions and that inoculation tools that are practically feasible for farmers are provided.

Other common agricultural crops are more difficult to inoculate at reasonable costs, mostly due to higher seeding rates or smaller yield gains after inoculation. Most studies in the field of AMF do not include an economic analysis and use laboratory cultures of AMF rather than commercial inoculants. Economic analysis on self-propagated cultures would be unreasonable as they are not produced on a commercial scale and therefore produced at much higher costs. Ignoring the economic factors of biofertilization, myriads of studies show that field inoculation with AMF is producing promising outcomes for plant growth (Thompson et al., 2013, Cely et al., 2016, Al-Karaki et al., 2004, Pellegrino and Bedini, 2014) and soil health (see Tables 2 and 3). In terms of efficient application of AMF for large seed quantities, seed coating is a promising solution to deliver not only AMF but also other beneficial microorganisms directly to the plant (Rocha et al., 2019b). Successful implementation of this technique, in combination with AMF and agricultural crops, has been reported (Rocha et al., 2019a, Oliveira et al., 2016), even in combination with fungicidal seed coating (Cameron et al., 2017).

4.2 Ecological restoration

The science of ecological restoration is focused on trajectories of change, that is, on the succession, assembly and state transition of natural communities. The

Table 2 Various meta-analysis on AMF involving their effect on plant growth and nutrient uptake. No additional analysis is presented here

Analyzed interaction	Outcome	Reference
Photosynthesis under salt stress	Alleviation of salinity stress and increased photosynthetic rate	Wang et al. (2019)
Wetland plant performance	Significant benefits, even under flood conditions	Ramírez-Viga et al. (2018)
Nutritional and non-nutritional factors in crops	Positive effects on nutrient uptake, soil aggregation, water flow and disease resistance	Delavaux et al. (2017)
Potato yield	Increased yields due to AMF inoculation, with economic benefits	Hijri (2016)
Response of wheat to AMF	AMF field inoculation increased nutrient uptake and dry weight	Pellegrino et al. (2015)
Copper, manganese and iron concentration in crops	Significant role for copper and iron and limited role for manganese	Lehmann and Rillig (2015)
Zinc nutrition in crop plants	Positive impacts on zinc concentration in shoot, root and fruit	Lehmann et al. (2014)
Nutrient uptake of tomato	Particularly beneficial for phosphorus and zinc	Watts-Williams and Cavagnaro (2014)
Plant growth under water stress	Improved drought resistance of plants	Jayne and Quigley (2014)
Mycorrhizal responsiveness in crop plants and wild relatives	No evidence of decreased mycorrhizal responsiveness	Lehmann et al. (2012)
Context-dependant plant response to AMF	Plant response most positive when plants are phosphorus limited	Hoeksema et al. (2010)

specific aims range from re-introduction of single species to population and community restorations. Broadly, ecological restoration can be defined as the assisted recovery of ecosystems that have been degraded or destroyed. Due to the various positive effects of AMF on the ecosystem and plants, they are considered a promising tool to drive this restoration process. In that way, they can increase the survivability of plant species in contaminated soils or favour certain plants over others, such as unwanted neophytes. Ultimately, restoration efforts in combination with AMF lead to the enhanced establishment of plant communities with positive effects on soil health (Asmelash et al., 2016).

In most cases, natural ecosystems do not allow active soil management such as in the case of agriculture. Most restoration efforts are focusing on transplanting and seeding of plants. Various studies demonstrated that AMF inoculation can be successfully merged with those methods. Zhang et al. (2012)

Table 3 Studies involving AMF field inoculation (biofertilization) and their effects on improving soil health

Host plant	Primary AMF ecosystem function	Outcome	Reference
Ricinus communis	Soil contamination	Increased plant survival and amelioration of lead soil pollution	González-Chávez et al. (2019)
Elymus trachycaulus, Lotus corniculatus	Soil contamination	Increased plant growth in coarse tailings sands	Boldt-Burisch et al. (2018)
Eight grassland plant species	Soil contamination	Higher plant cover in post-mine sandpits	Ohsowski et al. (2018)
Prunus discadenia, Prunus dictyneura, Xanthoceras sorbifolium, Armeniaca sibirica	Carbon sequestration, soil contamination	Increased carbon sequestration in coalfield soil	Wang et al. (2016)
Oryza glaberrima	Nutrient retention	Decreased nitrogen run-off at various fertilization levels	Zhang et al. (2016)
Viburnum tinus	Soil structure, soil contamination	Improved soil structure and growth of plants with wastewater treatment	Gómez-Bellot et al. (2015)
Oryza glaberrima	Nutrient retention	Decreased N_2O	Zhang et al. (2015)
Miscanthus × giganteus	Soil contamination	Alleviation of plant stress in metal contaminated sites	Firmin et al. (2015)
Agrostis capillaris	Soil contamination	Protective effect against arsenic toxicity in plants	Neagoe et al. (2014)
Meta-analysis	Soil structure	Overall positive effect of AMF inoculation on soil aggregation	Leifheit et al. (2014)
Soil macrocosm	Soil structure	Increased soil aggregate stability	Martin et al. (2012)
Erodium oxyrrhynchum, Hyalea pulchella, Trigonella arcuata, Schismus arabicus	Soil contamination	Increased plant cover and community productivity	Zhang et al. (2012)
Chrysopogon zizanioides	Soil contamination	Protective effects in lead and zinc mine tailings	Wu et al. (2010)
Medicago sativa	Soil contamination	Enhanced phytoremediation in polychlorinated biphenyl (PCB) contaminated soils	Teng et al. (2010)

Timeframe: 2020–2010.
Source: Scopus search: 'AMF mycorrhiza field inoculation'.

applied a mix of three lab-cultured AMF species to grassland by drilling holes and refilling them with the inoculum. Over 3 years, the number of established seedlings and their community productivity was significantly higher than in the uninoculated control. White et al. (2008) drilled seeds with inoculum into the soil, as well as broadcasting both onto the surface, achieving similar colonization results. However, in this case, there were no positive mycorrhizal effects due to the inoculation, which might be caused by high levels of phosphorus in the soil and a high abundance of native AMF communities. The efficiency between native and introduced AMF is still to be debated. Williams et al. (2012) pre-inoculated tree cuttings, whereas inoculum made of native AMFs and a natural forest performed better than a commercial one. The positive effects of seedling inoculation with natural AMF communities had also been demonstrated in other studies (Emam, 2016). It becomes evident that field inoculation of natural ecosystems with AMF is a feasible option for ecological restorations, and, as shown in some studies, can have long-lasting effects over multiple years. To this date, the species effectiveness between native and introduced AMF is not fully understood. However, we do have a good understanding of the importance that soil microorganisms have on above-ground ecological complexes. Managing below-ground mycorrhizas seems to be one important step in order to harbour the full potential of restoration efforts. Khan (2006) defined the term 'Mycorrhizoremediation' as an enhanced form of phytoremediation.

5 Conclusion

In this chapter we highlighted how AMF can enhance soil health and how they could be used as biofertilizers, either in agriculture or ecological restoration. The main mechanisms by which AMF improve soil health include improved soil structure and the immobilization of potentially toxic metals. The latest research even showed that AMF can enhance the degradation of organic pollutants. Due to the nature of their symbiosis, they act as a significant carbon sink for the soil, thereby increasing soil carbon. Carbon sequestration for soils is not only improving soil health but can also be used to trap atmospheric CO_2 back into the soil, thereby contributing to mitigate climate change. In the context of climate change, we also addressed the potential of AMF to reduce gaseous nutrient losses from soils, such as in the form of N_2O. This gas is mineralized from other forms of nitrogen in the soil and is a potent climate gas. AMF further reduce the leaching of nitrogen into the ground water and phosphorus-loss through soil erosion. Nutrient loss from soils is not only impacting soil fertility but is also critical with regard to the surrounding ecosystem, especially the aquatic biodiversity.

The highlighted effects of AMF on soil health can be used to increase sustainability in food production and to restore degraded soils. It is obvious that AMF could be applied as biofertilizers in various scenarios. Their economic and ecological potential was also commercialized by a variety of companies all around the world. The production and application of AMF is challenging and requires expertise. For people working in the field of mycorrhizal research, it is not surprising that many of those products fail to induce mycorrhizal colonization. This was also confirmed by some studies where only a fraction of the tested products were considered viable. During the past centuries, a variety of AMF production systems have been proposed which uses soil, hydroponics, aeroponics, or axenic cultures. Each system has its advantages and disadvantages, and to this date, the mass production of high-quality AMF propagules is an expensive endeavour. Nevertheless, investments in mycorrhizal products are fuelled by an endless stream of mycorrhizal research that highlights again and again the positive effects of AMF and how it could be used to reduce our footprint on this earth. The value of AMF field inoculation has been proven for a variety of scenarios, such as agriculture systems or to enhance phytoremediation efforts. In some cases where the targeted host plant is exceptionally responsive to mycorrhiza and where only relatively small amounts of inoculum are necessary, the use of AMF biofertilizer can be economically viable already today.

6 Future trends in research

In recent years, particularly with the advent of high throughput sequencing, our knowledge of soil microbial communities has increased greatly. It has long been known that soil microorganisms, including AMF, provide essential ecosystem services which we are now able to better quantify. The next great step in soil ecology involves the assignment of soil functions to microbial communities. In doing so, we will be able to quantify, and then predict, the impacts that management and other causes of environmental change (e.g. climate change) have on soils and the ecosystem services they provide. Armed with this knowledge we will then be able to better understand how to best manage systems to support the benefits of the soil microbiome. Such knowledge will also assist in the development of reliable biofertilizers. Those biofertilizers are not limited to AMF but the whole spectrum of beneficial soil microorganisms. One utopian vision of the future would involve a replacement of most agrochemicals with biofertilizers.

We have observed and also predict a continued interest in more sustainable farming systems in the future. Such systems seek to produce more food, on less land and with fewer inputs. At the same time, they are equipped against adverse conditions, such as the increasing occurrence of extreme weather events. This is a tall order but one we cannot afford to not meet in the context of an increasing global population and increased food requirements in the coming decades.

Moreover, we have observed a growing interest in agricultural paradigms that is shifting towards biologically regulated nutrient supply, rather than the importation of externally sourced synthetic inputs. This development would help to 'close the loop' and minimize the movement of resources on and off farm.

This is an exciting time to be studying soil ecology. There are many challenges in the here and now, and in the future to come. We contend that the soil microbiome, including mycorrhizal fungi, may hold many of the answers to meet those challenges and to do so in a sustainable way.

7 Where to look for further information

The following articles and books provide a good overview of the subject:

- Smith, S. E. and Read, D. (2008): Mycorrhizal Symbiosis, 3rd Edition, Academic Press, London.
- Brundrett, M., Bougher, N., Dell, B., Grove, T. and Malajczuk, N. (1995): Working with Mycorrhizas in Forestry and Agriculture, Australian Centre for International Agricultural Research, Canberra.
- Declerck, S., Strullu, D.-G. and Fortin, A. (2005): In Vitro Culture of Mycorrhizas, Springer, Berlin Heidelberg.
- Fisseha, A., Bekele, T. and Birhane, E. (2016): The Potential Role of Arbuscular Mycorrhizal Fungi in the Restoration of Degraded Lands, Front Microbiol. 7: 1095.

Research associations:

- International Mycorrhiza Society (http://mycorrhizas.org).

AMF collections:

- Banque Européenne des Glomeromycota (BEG) (www.kent.ac.uk/bio/beg).
- Glomeromycota In Vitro Collection (GINCO) (http://res2.agr.gc.ca/ecorc/ginco-can/index_e.htm).
- International Culture Collection of Vesicular-Arbuscular Mycorrhizal Fungi (INVAM) (http://invam.caf.wvu.edu).

8 References

Adhikari, K. and Hartemink, A. E. 2016. Linking soils to ecosystem services–a global review. *Geoderma* 262, 101-111.
Adholeya, A., Tiwari, P. and Singh, R. 2005. Large-scale inoculum production of arbuscular mycorrhizal fungi on root organs and inoculation strategies. In: Declerck, S., Fortin,

J. A. and Strullu, D.-G. (Eds) *In Vitro Culture of Mycorrhizas* (pp. 315–338). Berlin, Heidelberg: Springer.

Al-Karaki, G., McMichael, B. and Zak, J. 2004. Field response of wheat to arbuscular mycorrhizal fungi and drought stress. *Mycorrhiza* 14(4), 263–269.

Asmelash, F., Bekele, T. and Birhane, E. 2016. The potential role of arbuscular mycorrhizal fungi in the restoration of degraded lands. *Frontiers in Microbiology* 7, 1095.

Augé, R. M., Toler, H. D. and Saxton, A. M. 2015. Arbuscular mycorrhizal symbiosis alters stomatal conductance of host plants more under drought than under amply watered conditions: a meta-analysis. *Mycorrhiza* 25(1), 13–24.

Bender, S. F., Conen, F. and van der Heijden, M. G. A. 2015. Mycorrhizal effects on nutrient cycling, nutrient leaching and N2O production in experimental grassland. *Soil Biology and Biochemistry* 80, 283–292.

Bender, S. F., Plantenga, F., Neftel, A., Jocher, M., Oberholzer, H. R., Koehl, L., Giles, M., Daniell, T. J. and van der Heijden, M. G. A. 2014. Symbiotic relationships between soil fungi and plants reduce N_2O emissions from soil. *ISME Journal* 8(6), 1336–1345.

Bender, S. F., Schlaeppi, K., Held, A. and van der Heijden, M. G. A. 2019. Establishment success and crop growth effects of an arbuscular mycorrhizal fungus inoculated into Swiss corn fields. *Agriculture, Ecosystems and Environment* 273, 13–24.

Berruti, A., Lumini, E., Balestrini, R. and Bianciotto, V. 2016. Arbuscular mycorrhizal fungi as natural biofertilizers: let's benefit from past successes. *Frontiers in Microbiology* 6, 1559.

Boldt-Burisch, K., Naeth, M. A., Schneider, U., Schneider, B. and Hüttl, R. F. 2018. Plant growth and arbuscular mycorrhizae development in oil sands processing by-products. *Science of the Total Environment* 621, 30–39.

Borriello, R., Lumini, E., Girlanda, M., Bonfante, P. and Bianciotto, V. 2012. Effects of different management practices on arbuscular mycorrhizal fungal diversity in maize fields by a molecular approach. *Biology and Fertility of Soils* 48(8), 911–922.

Bowles, T. M., Barrios-Masias, F. H., Carlisle, E. A., Cavagnaro, T. R. and Jackson, L. E. 2016. Effects of arbuscular mycorrhizae on tomato yield, nutrient uptake, water relations, and soil carbon dynamics under deficit irrigation in field conditions. *Science of the Total Environment* 566–567, 1223–1234.

Bowles, T. M., Jackson, L. E., Loeher, M., Cavagnaro, T. R. and Nuñez, M. 2017. Ecological intensification and arbuscular mycorrhizas: a meta-analysis of tillage and cover crop effects. *Journal of Applied Ecology* 54(6), 1785–1793.

Bradley, R., Burt, A. J. and Read, D. J. 1981. Mycorrhizal infection and resistance to heavy metal toxicity in Calluna vulgaris. *Nature* 292(5821), 335–337.

Bünemann, E. K., Bongiorno, G., Bai, Z., Creamer, R. E., De Deyn, G., De Goede, R., Fleskens, L., Geissen, V., Kuyper, T. W., Mäder, P., Pulleman, M., Sukkel, W., Van Groenigen, J. W. and Brussaard, L. 2018. Soil quality—a critical review. *Soil Biology and Biochemistry* 120, 105–125.

Cameron, J. C., Lehman, R. M., Sexton, P., Osborne, S. L. and Taheri, W. I. 2017. Fungicidal seed coatings exert minor effects on arbuscular mycorrhizal fungi and plant nutrient content. *Agronomy Journal* 109(3), 1005–1012.

Cavagnaro, T. R., Bender, S. F., Asghari, H. R. and van der Heijden, M. G. A. 2015. The role of arbuscular mycorrhizas in reducing soil nutrient loss. *Trends in Plant Science* 20(5), 283–290.

Cely, M. V. T., De Oliveira, A. G., De Freitas, V. F., De Luca, M. B., Barazetti, A. R., Dos Santos, I. M. O., Gionco, B., Garcia, G. V., Prete, C. E. C. and Andrade, G. 2016. Inoculant of arbuscular mycorrhizal fungi (*Rhizophagus clarus*) increase yield of soybean and cotton under field conditions. *Frontiers in Microbiology* 7, 720.

Chaiyasen, A., Chaiya, L., Douds, D. D. and Lumyong, S. 2017. Influence of host plants and soil diluents on arbuscular mycorrhizal fungus propagation for on-farm inoculum production using leaf litter compost and agrowastes. *Biological Agriculture and Horticulture* 33(1), 52-62.

Chen, M., Arato, M., Borghi, L., Nouri, E. and Reinhardt, D. 2018. Beneficial services of arbuscular mycorrhizal fungi–from ecology to application. *Frontiers in Plant Science* 9, 1270.

Corkidi, L., Allen, E. B., Merhaut, D. J., Allen, M. F., Downer, J. A., Bohn, J. and Evans, M. 2017. Assessing the infectivity of commercial mycorrhizal inoculants in plant nursery conditions. *Journal of Environmental Horticulture* 22(3), 149-154.

Corkidi, L., Merhaut, D. J., Allen, E. B., Downer, J., Bohn, J. and Evans, M. 2011. Effects of mycorrhizal colonization on nitrogen and phosphorus leaching from nursery containers. *HortScience* 46(11), 1472-1479.

Declerck, S., Ijdo, M., Suarez, K. F., Voets, L. and De La Providencia, I. 2009. Method and system for in vitro mass production of arbuscular mycorrhizal fungi. Worldwide Patent Application.

Declerck, S., Strullu, D. G. and Plenchette, C. 1996. In vitro mass-production of the arbuscular mycorrhizal fungus, Glomus versiforme, associated with Ri T-DNA transformed carrot roots. *Mycological Research* 100(10), 1237-1242.

Del Val, C., Barea, J. M. and Azcón-Aguilar, C. 1999. Diversity of arbuscular mycorrhizal fungus populations in heavy-metal-contaminated soils. *Applied and Environmental Microbiology* 65(2), 718-723.

Delavaux, C. S., Smith-Ramesh, L. M. and Kuebbing, S. E. 2017. Beyond nutrients: a meta-analysis of the diverse effects of arbuscular mycorrhizal fungi on plants and soils. *Ecology* 98(8), 2111-2119.

Douds, D. and Shachar-Hill, Y. 2000. Carbon partitioning, cost, and metabolism of arbuscular mycorrhizas. In: Arbuscular Mycorrhizas: Physiology and Function (pp. 107-129). Springer, Dordrecht.

Douds, D. D., Pfeffer, P. E. and Shachar-Hill, Y. 2000. Application of in vitro methods to study carbon uptake and transport by AM fungi. *Plant and Soil* 226(2), 255-261.

Driver, J. D., Holben, W. E. and Rillig, M. C. 2005. Characterization of glomalin as a hyphal wall component of arbuscular mycorrhizal fungi. *Soil Biology and Biochemistry* 37(1), 101-106.

Elmes, R. P. and Mosse, B. 1984. Vesicular–arbuscular endomycorrhizal inoculum production. II. Experiments with maize (*Zea mays*) and other hosts in nutrient flow culture. *Canadian Journal of Botany* 62(7), 1531-1536.

Emam, T. 2016. Local soil, but not commercial AMF inoculum, increases native and non-native grass growth at a mine restoration site. *Restoration Ecology* 24(1), 35-44.

FAO 2015. Status of the world's soil resources (SWSR)-main report. *Food and Agriculture Organization of the United Nations and intergovernmental technical panel on soils*, Rome, Italy, 650. Available at: http://www.fao.org/documents/card/en/c/c6814873 -efc3-41db-b7d3-2081a10ede50/.

Firmin, S., Labidi, S., Fontaine, J., Laruelle, F., Tisserant, B., Nsanganwimana, F., Pourrut, B., Dalpé, Y., Grandmougin, A., Douay, F., Shirali, P., Verdin, A. and Lounès-Hadj Sahraoui, A. 2015. Arbuscular mycorrhizal fungal inoculation protects Miscanthus ×

giganteus against trace element toxicity in a highly metal-contaminated site. *Science of the Total Environment* 527, 91–99.

French, K. E. 2017. Engineering mycorrhizal symbioses to alter plant metabolism and improve crop health. *Frontiers in Microbiology* 8, 1403.

Galli, U., Schüepp, H. and Brunold, C. 1994. Heavy metal binding by mycorrhizal fungi. *Physiologia Plantarum* 92, 364–368.

Gebbink, M. F., Claessen, D., Bouma, B., Dijkhuizen, L. and Wösten, H. A. 2005. Amyloids–a functional coat for microorganisms. *Nature Reviews. Microbiology* 3(4), 333–341.

Gibbs, H. K. and Salmon, J. M. 2015. Mapping the world's degraded lands. *Applied Geography* 57, 12–21.

Göhre, V. and Paszkowski, U. 2006. Contribution of the arbuscular mycorrhizal symbiosis to heavy metal phytoremediation. *Planta* 223(6), 1115–1122.

Gómez-Bellot, M. J., Ortuño, M. F., Nortes, P. A., Vicente-Sánchez, J., Martín, F. F., Bañón, S. and Sánchez-Blanco, M. J. 2015. Protective effects of *Glomus iranicum var. tenuihypharum* on soil and *Viburnum tinus* plants irrigated with treated wastewater under field conditions. *Mycorrhiza* 25(5), 399–409.

González-Chávez, M. D. C. A., Carrillo-González, R., Cuellar-Sánchez, A., Delgado-Alvarado, A., Suárez-Espinosa, J., Ríos-Leal, E., Solís-Domínguez, F. A. and Maldonado-Mendoza, I. E. 2019. Phytoremediation assisted by mycorrhizal fungi of a Mexican defunct lead-acid battery recycling site. *Science of the Total Environment* 650(2), 3134–3144.

Hart, M. M., Antunes, P. M., Chaudhary, V. B. and Abbott, L. K. 2018. Fungal inoculants in the field: is the reward greater than the risk? *Functional Ecology* 32(1), 126–135.

Hawkins, H.-J. and George, E. 1997. Hydroponic culture of the mycorrhizal fungus *Glomus mosseae* with *Linum usitatissimum* L., Sorghum bicolor L. and *Triticum aestivum* L. *Plant and Soil* 196(1), 143–149.

Hijri, M. 2016. Analysis of a large dataset of mycorrhiza inoculation field trials on potato shows highly significant increases in yield. *Mycorrhiza* 26(3), 209–214.

Hildebrandt, U., Regvar, M. and Bothe, H. 2007. Arbuscular mycorrhiza and heavy metal tolerance. *Phytochemistry* 68(1), 139–146.

Hodge, A. and Fitter, A. H. 2010. Substantial nitrogen acquisition by arbuscular mycorrhizal fungi from organic material has implications for N cycling. *Proceedings of the National Academy of Sciences of the United States of America* 107(31), 13754–13759.

Hoeksema, J. D., Chaudhary, V. B., Gehring, C. A., Johnson, N. C., Karst, J., Koide, R. T., Pringle, A., Zabinski, C., Bever, J. D., Moore, J. C., Wilson, G. W. T., Klironomos, J. N. and Umbanhowar, J. 2010. A meta-analysis of context-dependency in plant response to inoculation with mycorrhizal fungi. *Ecology Letters* 13(3), 394–407.

Hol, W. H. G. and Cook, R. 2005. An overview of arbuscular mycorrhizal fungi–nematode interactions. *Basic and Applied Ecology* 6(6), 489–503.

Hung, L. L. and Sylvia, D. M. 1988. Production of vesicular-arbuscular mycorrhizal fungus inoculum in aeroponic culture. *Applied and Environmental Microbiology* 54(2), 353–357.

Ijdo, M., Cranenbrouck, S. and Declerck, S. 2011. Methods for large-scale production of AM fungi: past, present, and future. *Mycorrhiza* 21(1), 1–16.

Jarstfer, A. G. and Sylvia, D. M. 1995. Aeroponic culture of VAM fungi. In: Varma, A. and Hock, B. (Eds) *Mycorrhiza: Structure, Function, Molecular Biology and Biotechnology* (pp. 427–441). Berlin, Heidelberg: Springer.

Jayne, B. and Quigley, M. 2014. Influence of arbuscular mycorrhiza on growth and reproductive response of plants under water deficit: a meta-analysis. *Mycorrhiza* 24(2), 109–119.

Jeffries, P., Gianinazzi, S., Perotto, S., Turnau, K. and Barea, J.-M. 2003. The contribution of arbuscular mycorrhizal fungi in sustainable maintenance of plant health and soil fertility. *Biology and Fertility of Soils* 37(1), 1–16.

Jentschke, G., Brandes, B., Heinzemann, J., Marschner, P. and Godbold, D. L. 1999. Sand culture of mycorrhizal plants. *Journal of Plant Nutrition and Soil Science* 162(1), 107–112.

Jiang, Q. Y., Zhuo, F., Long, S. H., Zhao, H. D., Yang, D. J., Ye, Z. H., Li, S. S. and Jing, Y. X. 2016. Can arbuscular mycorrhizal fungi reduce Cd uptake and alleviate Cd toxicity of *Lonicera japonica* grown in Cd-added soils? *Scientific Reports* 6, 21805.

Jolicoeur, M., Williams, R. D., Chavarie, C., Fortin, J. A. and Archambault, J. 1999. Production of *Glomus intraradices* propagules, an arbuscular mycorrhizal fungus, in an airlift bioreactor. *Biotechnology and Bioengineering* 63(2), 224–232.

Kameoka, H., Tsutsui, I., Saito, K., Kikuchi, Y., Handa, Y., Ezawa, T., Hayashi, H., Kawaguchi, M. and Akiyama, K. 2019. Stimulation of asymbiotic sporulation in arbuscular mycorrhizal fungi by fatty acids. *Nature Microbiology* 4(10), 1654–1660.

Keesstra, S. D., Bouma, J., Wallinga, J., Tittonell, P., Smith, P., Cerdà, A., Montanarella, L., Quinton, J. N., Pachepsky, Y., Van Der Putten, W. H., Bardgett, R. D., Moolenaar, S., Mol, G., Jansen, B. and Fresco, L. O. 2016. The significance of soils and soil science towards realization of the United Nations sustainable development goals. *SOIL* 2(2), 111–128.

Khan, A. G. 2006. Mycorrhizoremediation—an enhanced form of phytoremediation. *Journal of Zhejiang University. Science. B* 7(7), 503–514.

Knowledge Sourcing Intelligence LLP 2019. Agricultural inoculants market—forecasts from 2019 to 2024. Available at: https://www.researchandmarkets.com/reports/48 95565/agricultural-inoculants-market-forecasts-from.

Kokkoris, V. and Hart, M. 2019. In vitro propagation of arbuscular mycorrhizal fungi may drive fungal evolution. *Frontiers in Microbiology* 10, 2420.

Kon Kam King, J., Granjou, C., Fournil, J. and Cecillon, L. 2018. Soil sciences and the French 4 per 1000 initiative—the promises of underground carbon. *Energy Research and Social Science* 45, 144–152.

Lanfranco, L. 2007. The fine-tuning of heavy metals in mycorrhizal fungi. *New Phytologist* 174(1), 3–6.

Lazcano, C., Barrios-Masias, F. H. and Jackson, L. E. 2014. Arbuscular mycorrhizal effects on plant water relations and soil greenhouse gas emissions under changing moisture regimes. *Soil Biology and Biochemistry* 74, 184–192.

Lehmann, A., Barto, E. K., Powell, J. R. and Rillig, M. C. 2012. Mycorrhizal responsiveness trends in annual crop plants and their wild relatives—a meta-analysis on studies from 1981 to 2010. *Plant and Soil* 355(1–2), 231–250.

Lehmann, A. and Rillig, M. C. 2015. Arbuscular mycorrhizal contribution to copper, manganese and iron nutrient concentrations in crops—a meta-analysis. *Soil Biology and Biochemistry* 81, 147–158.

Lehmann, A., Veresoglou, S. D., Leifheit, E. F. and Rillig, M. C. 2014. Arbuscular mycorrhizal influence on zinc nutrition in crop plants—a meta-analysis. *Soil Biology and Biochemistry* 69, 123–131.

Leifheit, E. F., Veresoglou, S. D., Lehmann, A., Morris, E. K. and Rillig, M. C. 2014. Multiple factors influence the role of arbuscular mycorrhizal fungi in soil aggregation–a meta-analysis. *Plant and Soil* 374(1–2), 523–537.

Martin, S. L., Mooney, S. J., Dickinson, M. J. and West, H. M. 2012. Soil structural responses to alterations in soil microbiota induced by the dilution method and mycorrhizal fungal inoculation. *Pedobiologia* 55(5), 271–281.

Miller, R. M. and Jastrow, J. D. 1990. Hierarchy of root and mycorrhizal fungal interactions with soil aggregation. *Soil Biology and Biochemistry* 22(5), 579–584.

Miller, R. M. and Jastrow, J. D. 2000. Mycorrhizal fungi influence soil structure. In: Kapulnik, Y. and Douds, D. D. (Eds) *Arbuscular Mycorrhizas: Physiology and Function* (pp. 3–18). Dordrecht, Netherlands: Springer.

Munkholm, L. J., Heck, R. J. and Deen, B. 2013. Long-term rotation and tillage effects on soil structure and crop yield. *Soil and Tillage Research* 127, 85–91.

Neagoe, A., Stancu, P., Nicoară, A., Onete, M., Bodescu, F., Gheorghe, R. and Iordache, V. 2014. Effects of arbuscular mycorrhizal fungi on *Agrostis capillaris* grown on amended mine tailing substrate at pot, lysimeter, and field plot scales. *Environmental Science and Pollution Research International* 21(11), 6859–6876.

Oehl, F., Sieverding, E., Mader, P., Dubois, D., Ineichen, K., Boller, T. and Wiemken, A. 2004. Impact of long-term conventional and organic farming on the diversity of arbuscular mycorrhizal fungi. *Oecologia* 138(4), 574–583.

Oliveira, R. S., Rocha, I., Ma, Y., Vosátka, M. and Freitas, H. 2016. Seed coating with arbuscular mycorrhizal fungi as an ecotechnological approach for sustainable agricultural production of common wheat (Triticum aestivum L.). *Journal of Toxicology and Environmental Health. Part A* 79(7), 329–337.

Olsson, L., Barbosa, H., Bhadwal, S., Cowie, A., Delusca, K., Flores-Renteria, D., Hermans, K., Jobbagy, E., Kurz, W., Li, D., Sonwa, D. J. and Stringer, L. 2019. Land Degradation: IPCC Special Report on Climate Change, Desertification, Land 5 Degradation, Sustainable Land Management, Food Security, and 6 Greenhouse gas fluxes in Terrestrial Ecosystems. Available at: https://www.ipcc.ch/srccl/chapter/chapter-4/.

Ohsowski, B. M., Dunfield, K., Klironomos, J. N. and Hart, M. M. 2018. Plant response to biochar, compost, and mycorrhizal fungal amendments in post-mine sandpits. *Restoration Ecology* 26(1), 63–72.

Pellegrino, E. and Bedini, S. 2014. Enhancing ecosystem services in sustainable agriculture: biofertilization and biofortification of chickpea (*Cicer arietinum* L.) by arbuscular mycorrhizal fungi. *Soil Biology and Biochemistry* 68, 429–439.

Pellegrino, E., Öpik, M., Bonari, E. and Ercoli, L. 2015. Responses of wheat to arbuscular mycorrhizal fungi: a meta-analysis of field studies from 1975 to 2013. *Soil Biology and Biochemistry* 84, 210–217.

Pérès, G., Cluzeau, D., Menasseri, S., Soussana, J. F., Bessler, H., Engels, C., Habekost, M., Gleixner, G., Weigelt, A., Weisser, W. W., Scheu, S. and Eisenhauer, N. 2013. Mechanisms linking plant community properties to soil aggregate stability in an experimental grassland plant diversity gradient. *Plant and Soil* 373(1–2), 285–299.

Peterson, R. L. and Massicotte, H. B. 2004. Exploring structural definitions of mycorrhizas, with emphasis on nutrient-exchange interfaces. Canadian Journal of Botany 82(8), 1074–1088.

Piotrowski, J. S., Denich, T., Klironomos, J. N., Graham, J. M. and Rillig, M. C. 2004. The effects of arbuscular mycorrhizas on soil aggregation depend on the interaction between plant and fungal species. *New Phytologist* 164(2), 365–373.

Pozo, M. J. and Azcón-Aguilar, C. 2007. Unraveling mycorrhiza-induced resistance. *Current Opinion in Plant Biology* 10(4), 393–398.

Quan, Z., Li, S., Zhang, X., Zhu, F., Li, P., Sheng, R., Chen, X., Zhang, L.-M., He, J.-Z., Wei, W. and Fang, Y. 2020. Fertilizer nitrogen use efficiency and fates in maize cropping systems across China: field 15N tracer studies. *Soil and Tillage Research* 197, 104498.

Ramírez-Viga, T. K., Aguilar, R., Castillo-Argüero, S., Chiappa-Carrara, X., Guadarrama, P. and Ramos-Zapata, J. 2018. Wetland plant species improve performance when inoculated with arbuscular mycorrhizal fungi: a meta-analysis of experimental pot studies. *Mycorrhiza* 28(5–6), 477–493.

Remy, W., Taylor, T. N., Hass, H. and Kerp, H. 1994. Four hundred-million-year-old vesicular arbuscular mycorrhizae. *Proceedings of the National Academy of Sciences of the United States of America* 91(25), 11841–11843.

Rillig, M. C. 2005. A connection between fungal hydrophobins and soil water repellency? *Pedobiologia* 49(5), 395–399.

Rillig, M. C., Aguilar-Trigueros, C. A., Camenzind, T., Cavagnaro, T. R., Degrune, F., Hohmann, P., Lammel, D. R., Mansour, I., Roy, J., van der Heijden, M. G. A. and Yang, G. 2019a. Why farmers should manage the arbuscular mycorrhizal symbiosis. *New Phytologist* 222(3), 1171–1175.

Rillig, M. C., Ryo, M., Lehmann, A., Aguilar-Trigueros, C. A., Buchert, S., Wulf, A., Iwasaki, A., Roy, J. and Yang, G. 2019b. The role of multiple global change factors in driving soil functions and microbial biodiversity. *Science* 366(6467), 886–890.

Rillig, M. C. and Mummey, D. L. 2006. Mycorrhizas and soil structure. *New Phytologist* 171(1), 41–53.

Rivero, J., Álvarez, D., Flors, V., Azcón-Aguilar, C. and Pozo, M. J. 2018. Root metabolic plasticity underlies functional diversity in mycorrhiza-enhanced stress tolerance in tomato. *New Phytologist* 220(4), 1322–1336.

Rocha, I., Ma, Y., Carvalho, M. F., Magalhães, C., Janoušková, M., Vosátka, M., Freitas, H. and Oliveira, R. S. 2019a. Seed coating with inocula of arbuscular mycorrhizal fungi and plant growth promoting rhizobacteria for nutritional enhancement of maize under different fertilisation regimes. *Archives of Agronomy and Soil Science* 65(1), 31–43.

Rocha, I., Ma, Y., Souza-Alonso, P., Vosátka, M., Freitas, H. and Oliveira, R. S. 2019b. Seed coating: a tool for delivering beneficial microbes to agricultural crops. *Frontiers in Plant Science* 10, 1357.

Rockström, J., Steffen, W., Noone, K., Persson, A., Chapin, F. S., Lambin, E. F., Lenton, T. M., Scheffer, M., Folke, C., Schellnhuber, H. J., Nykvist, B., De Wit, C. A., Hughes, T., Van Der Leeuw, S., Rodhe, H., Sörlin, S., Snyder, P. K., Costanza, R., Svedin, U., Falkenmark, M., Karlberg, L., Corell, R. W., Fabry, V. J., Hansen, J., Walker, B., Liverman, D., Richardson, K., Crutzen, P. and Foley, J. A. 2009. A safe operating space for humanity. *Nature* 461(7263), 472–475.

Rodriguez, A. and Sanders, I. R. 2015. The role of community and population ecology in applying mycorrhizal fungi for improved food security. *The ISME Journal* 9(5), 1053–1061.

Rosier, C. L., Hoye, A. T. and Rillig, M. C. 2006. Glomalin-related soil protein: assessment of current detection and quantification tools. *Soil Biology and Biochemistry* 38(8), 2205–2211.

Rouphael, Y., Franken, P., Schneider, C., Schwarz, D., Giovannetti, M., Agnolucci, M., Pascale, S. D., Bonini, P. and Colla, G. 2015. Arbuscular mycorrhizal fungi act as biostimulants in horticultural crops. *Scientia Horticulturae* 196, 91–108.

Rumpel, C., Amiraslani, F., Chenu, C., Garcia Cardenas, M., Kaonga, M., Koutika, L. S., Ladha, J., Madari, B., Shirato, Y., Smith, P., Soudi, B., Soussana, J. F., Whitehead, D. and Wollenberg, E. 2020. The 4p1000 initiative: opportunities, limitations and challenges for implementing soil organic carbon sequestration as a sustainable development strategy. *Ambio* 49(1), 350-360.

Ryan, M. H. and Graham, J. H. 2018. Little evidence that farmers should consider abundance or diversity of arbuscular mycorrhizal fungi when managing crops. *New Phytologist* 220(4), 1092-1107.

Salomon, M. J., Watts-Williams, S. J., McLaughlin, M. J. and Cavagnaro, T. R. 2020. Urban soil health: a city-wide survey of chemical and biological properties of urban agriculture soils. *Journal of Cleaner Production* 275, 122900.

Sanchez-Diaz, M. and Honrubia, M. 1994. Water relations and alleviation of drought stress in mycorrhizal plants. In: *Impact of Arbuscular Mycorrhizas on Sustainable Agriculture and Natural Ecosystems* (pp. 167-178). Basel: Birkhäuser.

Sanderman, J., Hengl, T. and Fiske, G. J. 2017. Soil carbon debt of 12,000 years of human land use. *Proceedings of the National Academy of Sciences* 114(36), 9575 -9580.

Sanders, I. R. 2010. 'Designer' mycorrhizas?: using natural genetic variation in AM fungi to increase plant growth. *The ISME Journal* 4(9), 1081-1083.

Scharlemann, J. P. W., Tanner, E. V. J., Hiederer, R. and Kapos, V. 2014. Global soil carbon: understanding and managing the largest terrestrial carbon pool. *Carbon Management* 5(1), 81-91.

Six, J., Bossuyt, H., Degryze, S. and Denef, K. 2004. A history of research on the link between (micro) aggregates, soil biota, and soil organic matter dynamics. *Soil and Tillage Research* 79(1), 7-31.

Smith, S. E. and Read, D. 2008. Mycorrhizal symbiosis. In: Smith, S. E. and Read, D. (Eds) *Mycorrhizal Symbiosis* (3rd edn., pp. 11-145). London: Academic Press.

Smith, S. E., Smith, F. A. and Jakobsen, I. 2004. Functional diversity in arbuscular mycorrhizal (AM) symbioses: the contribution of the mycorrhizal P uptake pathway is not correlated with mycorrhizal responses in growth or total P uptake. *New Phytologist* 162(2), 511-524.

Soudzilovskaia, N. A., van der Heijden, M. G. A., Cornelissen, J. H. C., Makarov, M. I., Onipchenko, V. G., Maslov, M. N., Akhmetzhanova, A. A. and Van Bodegom, P. M. 2015. Quantitative assessment of the differential impacts of arbuscular and ectomycorrhiza on soil carbon cycling. *New Phytologist* 208(1), 280-293.

Syers, J., Johnston, A. and Curtin, D. 2008. Efficiency of soil and fertilizer phosphorus use. *FAO Fertilizer and Plant Nutrition Bulletin* 18. Rome, Italy: Food and Agriculture Organization of the United Nations.

Sylvia, D. M. and Jarstfer, A. G. 1992. Sheared-root inocula of vesicular-arbuscular mycorrhizal fungi. *Applied and Environmental Microbiology* 58(1), 229-232.

Tan, Z. X., Lal, R. and Wiebe, K. D. 2005. Global soil nutrient depletion and yield reduction. *Journal of Sustainable Agriculture* 26(1), 123-146.

Tarbell, T. J. and Koske, R. E. 2007. Evaluation of commercial arbuscular mycorrhizal inocula in a sand/peat medium. *Mycorrhiza* 18(1), 51-56.

Tawaraya, K., Hirose, R. and Wagatsuma, T. 2012. Inoculation of arbuscular mycorrhizal fungi can substantially reduce phosphate fertilizer application to Allium fistulosum L. and achieve marketable yield under field condition. *Biology and Fertility of Soils* 48(7), 839-843.

Taylor, L. L., Leake, J. R., Quirk, J., Hardy, K., Banwart, S. A. and Beerling, D. J. 2009. Biological weathering and the long-term carbon cycle: integrating mycorrhizal evolution and function into the current paradigm. *Geobiology* 7(2), 171-191.

Teng, Y., Luo, Y., Sun, X., Tu, C., Xu, L., Liu, W., Li, Z. and Christie, P. 2010. Influence of arbuscular mycorrhiza and rhizobium on phytoremediation by alfalfa of an agricultural soil contaminated with weathered PCbs: a field study. *International Journal of Phytoremediation* 12(5), 516-533.

Thompson, J. P., Clewett, T. G. and Fiske, M. L. 2013. Field inoculation with arbuscular-mycorrhizal fungi overcomes phosphorus and zinc deficiencies of linseed (Linum usitatissimum) in a vertisol subject to long-fallow disorder. *Plant and Soil* 371(1-2), 117-137.

Tisdall, J. 1991. Fungal hyphae and structural stability of soil. *Soil Research* 29(6), 729-743.

Tisdall, J. M. and Oades, J. M. 1982. Organic matter and water-stable aggregates in soils. *Journal of Soil Science* 33(2), 141-163.

Valetti, L., Iriarte, L. and Fabra, A. 2016. Effect of previous cropping of rapeseed (Brassica napus L.) on soybean (Glycine max) root mycorrhization, nodulation, and plant growth. *European Journal of Soil Biology* 76, 103-106.

van der Heijden, M. G. 2010. Mycorrhizal fungi reduce nutrient loss from model grassland ecosystems. *Ecology* 91(4), 1163-1171.

van der Heijden, M. G. A., Bardgett, R. D. and Van Straalen, N. M. 2008. The unseen majority: soil microbes as drivers of plant diversity and productivity in terrestrial ecosystems. *Ecology Letters* 11(3), 296-310.

van der Heijden, M. G. A., Klironomos, J. N., Ursic, M., Moutoglis, P., Streitwolf-Engel, R., Boller, T., Wiemken, A. and Sanders, I. R. 1998. Mycorrhizal fungal diversity determines plant biodiversity, ecosystem variability and productivity. *Nature* 396(6706), 69-72.

van der Heijden, M. G. A., Streitwolf-Engel, R., Riedl, R., Siegrist, S., Neudecker, A., Ineichen, K., Boller, T., Wiemken, A. and Sanders, I. R. 2006. The mycorrhizal contribution to plant productivity, plant nutrition and soil structure in experimental grassland. *New Phytologist* 172(4), 739-752.

Verbruggen, E., Röling, W. F. M., Gamper, H. A., Kowalchuk, G. A., Verhoef, H. A. and van der Heijden, M. G. A. 2010. Positive effects of organic farming on below-ground mutualists: large-scale comparison of mycorrhizal fungal communities in agricultural soils. *New Phytologist* 186(4), 968-979.

Voets, L., Dupré De Boulois, H., Renard, L., Strullu, D. G. and Declerck, S. 2005. Development of an autotrophic culture system for the in vitro mycorrhization of potato plantlets. *FEMS Microbiology Letters* 248(1), 111-118.

Vogel-Mikuš, K., Drobne, D. and Regvar, M. 2005. Zn, Cd and Pb accumulation and arbuscular mycorrhizal colonisation of pennycress *Thlaspi praecox* Wulf. (Brassicaceae) from the vicinity of a lead mine and smelter in Slovenia. *Environmental Pollution* 133(2), 233-242.

Vogel-Mikuš, K. and Regvar, M. 2006. Arbuscular mycorrhiza as a tolerance strategy in metal contaminated soils: prospects in phytoremediation. *New Topics in Environmental Research*, 37-56.

Vosátka, M., Albrechtová, J. and Patten, R. 2008. The international market development for mycorrhizal technology. In: Varma, A. (Ed) *Mycorrhiza: State of the Art, Genetics and Molecular Biology, Eco-Function, Biotechnology, Eco-Physiology, Structure and Systematics* (pp. 419-438). Berlin, Heidelberg: Springer.

Vosátka, M., Látr, A., Gianinazzi, S. and Albrechtová, J. 2012. Development of arbuscular mycorrhizal biotechnology and industry: current achievements and bottlenecks. *Symbiosis* 58(1-3), 29-37.

Walker, C. and Vestberg, M. 1994. A simple and inexpensive method for producing and maintaining closed pot cultures of arbuscular mycorrhizal fungi. *Agricultural and Food Science* 3(3), 233-240.

Wallander, H., Nilsson, L. O., Hagerberg, D. and Bååth, E. 2001. Estimation of the biomass and seasonal growth of external mycelium of ectomycorrhizal fungi in the field. *New Phytologist* 151(3), 753-760.

Wang, F., Adams, C. A., Yang, W., Sun, Y. and Shi, Z. 2020. Benefits of arbuscular mycorrhizal fungi in reducing organic contaminant residues in crops: implications for cleaner agricultural production. *Critical Reviews in Environmental Science and Technology* 50(15), 1580-1612.

Wang, Y., Wang, J., Yan, X., Sun, S. and Lin, J. 2019. The effect of arbuscular mycorrhizal fungi on photosystem II of the host plant Under salt stress: a meta-analysis. *Agronomy* 9(12), 806.

Wang, Z. G., Bi, Y. L., Jiang, B., Zhakypbek, Y., Peng, S. P., Liu, W. W. and Liu, H. 2016. Arbuscular mycorrhizal fungi enhance soil carbon sequestration in the coalfields, northwest China. *Scientific Reports* 6, 34336.

Watts-Williams, S. J. and Cavagnaro, T. R. 2012. Arbuscular mycorrhizas modify tomato responses to soil zinc and phosphorus addition. *Biology and Fertility of Soils* 48(3), 285-294.

Watts-Williams, S. J. and Cavagnaro, T. R. 2014. Nutrient interactions and arbuscular mycorrhizas: a meta-analysis of a mycorrhiza-defective mutant and wild-type tomato genotype pair. *Plant and Soil* 384(1-2), 79-92.

White, J. A., Tallaksen, J. and Charvat, I. 2008. The effects of arbuscular mycorrhizal fungal inoculation at a roadside prairie restoration site. *Mycologia* 100(1), 6-11.

Williams, A., Norton, D. A. and Ridgway, H. J. 2012. Different arbuscular mycorrhizal inoculants affect the growth and survival of *Podocarpus cunninghamii* restoration plantings in the Mackenzie Basin, New Zealand. *New Zealand Journal of Botany* 50(4), 473-479.

Wilson, G. W., Rice, C. W., Rillig, M. C., Springer, A. and Hartnett, D. C. 2009. Soil aggregation and carbon sequestration are tightly correlated with the abundance of arbuscular mycorrhizal fungi: results from long-term field experiments. *Ecology Letters* 12(5), 452-461.

Wu, S. C., Wong, C. C., Shu, W. S., Khan, A. G. and Wong, M. H. 2010. Mycorrhizo-remediation of lead/zinc mine tailings using vetiver: a field study. *International Journal of Phytoremediation* 13(1), 61-74.

Wuana, R. A. and Okieimen, F. E. 2011. Heavy metals in contaminated soils: a review of sources, chemistry, risks and best available strategies for remediation. *ISRN Ecology* 2011, 1-20, 402647.

Yang, Y., He, C., Huang, L., Ban, Y. and Tang, M. 2017. The effects of arbuscular mycorrhizal fungi on glomalin-related soil protein distribution, aggregate stability and their relationships with soil properties at different soil depths in lead-zinc contaminated area. *PLoS ONE* 12(8), e0182264.

Zhang, S., Wang, L., Ma, F., Zhang, X. and Fu, D. 2016. Reducing nitrogen runoff from paddy fields with arbuscular mycorrhizal fungi under different fertilizer regimes. *Journal of Environmental Sciences* 46, 92-100.

Zhang, T., Sun, Y., Shi, Z. and Feng, G. 2012. Arbuscular mycorrhizal fungi can accelerate the restoration of degraded spring grassland in Central Asia. *Rangeland Ecology and Management* 65(4), 426–432.

Zhang, X., Wang, L., Ma, F. and Shan, D. 2015. Effects of arbuscular mycorrhizal fungi on N20 emissions from rice paddies. *Water, Air, and Soil Pollution* 226(7), 222.

Zhu, Y. G. and Michael Miller, R. M. 2003. Carbon cycling by arbuscular mycorrhizal fungi in soil-plant systems. *Trends in Plant Science* 8(9), 407–409.

Chapter 5

Biofertilizers: assessing the effects of plant growth-promoting bacteria (PGPB) or rhizobacteria (PGPR) on soil and plant health

Elisa Zampieri, Institute for Sustainable Plant Protection, Italy; Iakovos S. Pantelides, Cyprus University of Technology, Cyprus; and Raffaella Balestrini, Institute for Sustainable Plant Protection, Italy

1 Introduction

The green revolution was driven by agricultural intensification resulting in increased productivity and incomes, but also a dependence on chemical substances in many developing countries of the world (Aeron et al., 2020). The uncontrolled application of synthetic agrochemicals imposes serious negative impacts on the environment and human and animal health, leading to a reduction in soil fertility and microbial diversity, soil pollution and environmental degradation and the development of resistance in phytopathogens and pests (Aeron et al., 2020). However, global demand for agricultural crops is increasing with yields still insufficient to face the ever-growing food demand (Timmusk et al., 2017). In this context, unsustainable agricultural intensification has often led to pollution, overexploitation of natural areas and resources, loss of soil fertility, soil erosion, salinization, runoff and in some cases desertification (IPCC, 2019, special

http://dx.doi.org/10.19103/AS.2021.0094.18

report; https://www.ipcc.ch/srccl/). Drought and land degradation following the salinization of soil are considerably increasing worldwide, and the ongoing climate change could amplify the negative effects of this scenario (Corwin, 2021). For these reasons, together with the increasing awareness of consumers about healthy food, sustainable agricultural practices are encouraged as alternatives to mineral fertilizers and synthetic pesticides (Brodt et al., 2011).

Sustainable agricultural management practices include the use of beneficial microorganisms, such as mycorrhizal fungi, rhizobia and other plant growth-promoting bacteria (PGPB) or rhizobacteria (PGPR), to support plant protection and nutrition and assist water conservation. Today, these beneficial microorganisms (i.e. arbuscular mycorrhizal fungi (AMF), PGPB or PGPR) are considered a key factor for managing crop productions (Schlaeppi and Bulgarelli, 2015). However, their application in agriculture is still a challenge due to inconsistent and unpredictable results, which often are context-dependent (dos Santos et al., 2020; Compant et al., 2019). There are many aspects that need to be considered for a successful implementation of AMF and PGPB/PGPR as microbial inoculants with desired outputs in different crop genotypes and upon different (and combined) stress conditions (Pascale et al., 2020). These microbial inoculants are living microorganisms that colonize the rhizosphere (i.e. the zone surrounding the roots that is directly influenced by plant root secretions) and/or the inner regions of plant tissues and promote plant growth or act as biological control agents (BCAs) against soilborne and seedborne plant pathogens (Aeron et al., 2020; Khatoon et al., 2020; Raj et al., 2020; Tsolakidou et al., 2019; Orozco-Mosqueda et al., 2018; Bhattacharyya and Jha, 2012). Additionally, understanding the effect of cropland management on soil microorganism dynamics is fundamental for designing better management practices to restore soil function in intensively managed agricultural systems (Baritz et al., 2018; http://www.fao.org/3/a-bl813e.pdf).

2 Mechanisms mediated by plant growth-promoting bacteria/rhizobacteria

Generally, about 2-5% of the total rhizospheric bacteria are PGPB/PGPR (Antoun and Prévost, 2006). Features that allow bacterial survival in the rhizosphere and plant tissue colonization are motility, chemotaxis, attachment, growth and stress resistance (Bulgarelli et al., 2013). Some PGPB/PGPR are considered biofertilizers that augment the availability of nutrients in a form that can be easily assimilated by plants, while others act as biocontrol agents or biopesticides that suppress or control plant disease (Timmusk et al., 2017). Many PGPB/PGPR can solubilize insoluble soil phosphate to release soluble phosphorus and making it available to plants. This trait is very interesting since the phosphorus content of soil is about 0.05% (w/w) but only 0.1% of this fraction is available to plants. Phosphorus is an essential element involved in many important metabolic pathways, such as photosynthesis, respiration, electron transport chain, biosynthesis of macromolecules and signal transduction (Khan et al., 2010). It also influences root growth, seed development and normal crop maturity (Heydari et al., 2019). Many bacterial and fungal strains, such as *Bacillus*, *Pseudomonas* or *Penicillium*, that release organic acids or phosphatases are capable to solubilize phosphorus and are, therefore, promising as PGPR (Khatoon et al., 2020; Bulgarelli et al., 2013). Apart from phosphate solubilisation (Figure 1), many other mechanisms mediated by PGPB/PGPR can lead to plant-growth promotion and improve plant tolerance/resistance to abiotic and biotic stresses (Glick, 2012), such as synthesis of hormones (abscisic acid (ABA), gibberellic acid, cytokinins and auxins) (Pérez-Flores et al., 2017; Bhattacharyya et al., 2015), nitrogen fixation (Ashraf et al., 2011), solubilization of other nutrients (Zn, K) (Vikram and Hamzehzarghani, 2008; Etesami et al., 2017; Zaheer et al., 2019), production of siderophores (Sinha and Parli, 2020; Calvo et al., 2017), ethylene (ET) control in emerging plants under stress conditions through the production of aminocyclopropane-1-carboxylate (ACC) deaminase (Ravanbakhsh et al., 2018; Glick, 2012), secretion of several molecules, including antibiotics, hydrolytic enzymes and volatile organic compounds, alleviating biotic stress effects and contributing to systemic resistance (Meena et al., 2020; Kour et al., 2019; Orozco-Mosqueda et al., 2018), production of exopolysaccharides (EPS) and biofilm formation (Dimkpa et al., 2009), heavy metal detoxification (Sharma and Archana, 2016; Tak et al., 2013; Ma et al., 2011) and pest management (Subbanna et al., 2018).

An increasing number of plants have been reported to benefit from PGPB/PGPR presence (Santos et al., 2019), including the model plant *Arabidopsis thaliana* (Lee et al., 2020) and several crops, such as winter wheat (*Triticum aestivum*) (Awan et al., 2020; Turan et al., 2012) and wheat (*Triticum durum*) (Bechtaoui et al., 2019), rice (*Oryza sativa*) (Xiao et al., 2020; Andreozzi et al., 2019;), sunflower (*Helianthus annuus*) (Ambrosini et al., 2012), rape (*Brassica napus*) (Farina et al., 2012), runner bean (*Phaseolus coccineus*) (Stefan et al., 2013) and faba bean (*Vicia faba*) (Bechtaoui et al., 2019), corn (*Zea mays*) (Batista et al., 2018; Tchuisseu Tchakounté et al., 2018; Arruda et al., 2013), soybean (*Glycine max*) (Batista et al., 2018), chickpea (*Cicer arietinum*) (Bisht et al., 2019), tomato (*Solanum lycopersicum*) (Kalam et al., 2020; Pellegrini et al., 2020), potato (*Solanum tuberosum*) (Pellegrini et al., 2020), flax (*Linum usitatissimum*) (Planchon et al., 2021), coriander (*Coriandrum sativum*) (Jiménez-Gómez, et al., 2020) and spinach (*Spinacia oleracea*) (Zafar-Ul-Hye et al., 2020). Apart from increased plant biomass, PGPR have demonstrated positive effects on total oil content and lipid composition in *G. max*, *B. napus* and *Buglossoides arvensis* that are important sources of oleic, linoleic, α-linolenic and omega-3 stearidonic acids (Jiménez et al., 2020). In addition, PGPR have been reported to improve carotenoids, tocopherols, and folate content in horseradish (*Moringa oleifera*) (Sonbarse et al, 2020).

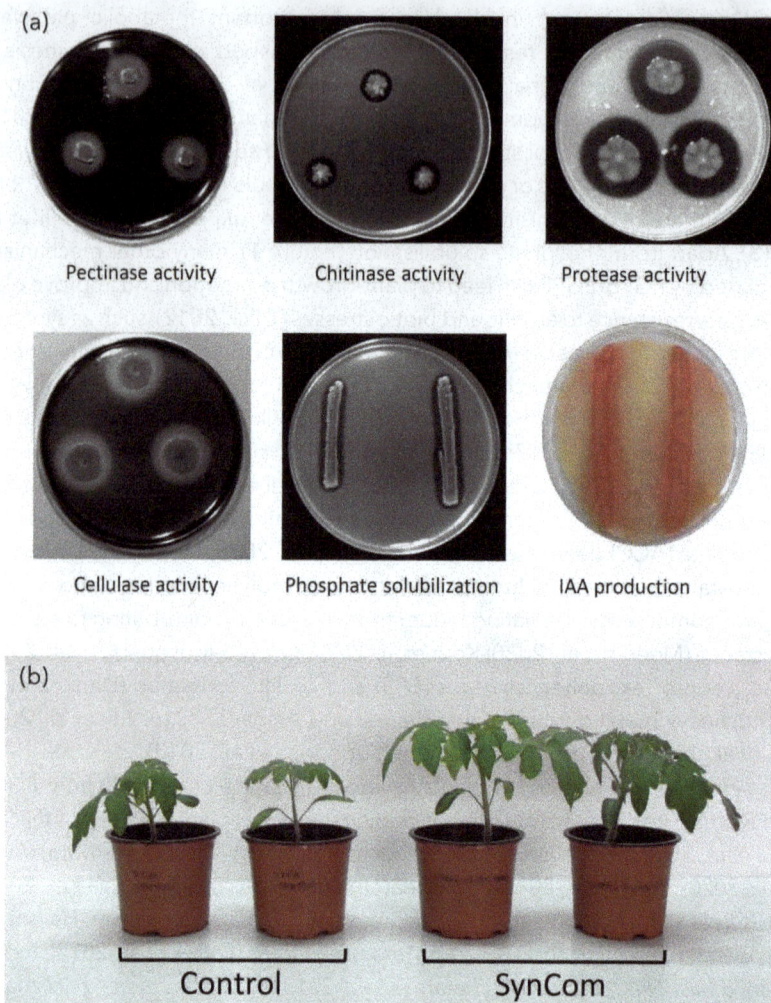

Figure 1 (a) *In vitro* screening of plant growth promotion traits; (b) Tomato plants not inoculated (left) and inoculated (right) with a microbial synthetic community (SynCom). [Photographs by I. S. Pantelides, Cyprus University of Technology].

PGPR have also been known to mediate biotic stress tolerance in plants through the production of antimicrobial compounds and the induction of plant defence responses. Ali et al. (2020) recently isolated bacteria from maize, rice, wheat, potato, sunflower and soybean rhizosphere and verified the antifungal activity against *Fusarium oxysporum*, *Fusarium moniliforme*, *Rhizoctonia solani*, *Colletotrichum gloeosporioides*, *Colletotrichum falcatum*, *Aspergillus niger* and *Aspergillus flavus*. The PGPR showing the highest antagonistic activity belonged to *Pseudomonas* and *Bacillus* species (Ali et al., 2020). The same genera have also suppressed *Phytophthora capsici* infections in chilli pepper (Hyder et al., 2020). Multiple strains of *Bacillus* spp. together with a strain of *Stenotrophomonas rhizophila* were also effective in reducing *Meloidogyne incognita* population density and improving turfgrass root growth (Groover et al., 2020). In addition, *Bacillus amyloliquefaciens* strain S1 exhibited high *in vitro* antagonistic activity against *Clavibacter michiganensis* ssp. *michiganensis,* suggesting its possible employment in controlling bacterial canker in tomato plants (Gautam et al., 2019).

Following a transcriptomics approach, Gamez et al. (2019) highlighted that PGPR inoculation in banana (*Musa acuminata* Colla) cv. Williams resulted in differential expression of genes related to growth promotion and regulation of specific functions (flowering, photosynthesis, glucose catabolism and root growth) as well as genes involved in plant defence. Jiang et al. (2019) also demonstrated that the watermelon gene expression profile was altered in the presence of a *Bacillus* strain in combination with *F. oxysporum* f. sp. *niveum.* The *Bacillus* strain enhanced plant disease resistance against the pathogen through activation of defence-related genes and phytohormone signal factors (Jiang et al., 2019). Recently, Bertani et al. (2021) showed the expression of rice genes involved in ET and auxin pathways together with genes coding for a metallothionein-like protein and a multiple stress-responsive zinc-finger protein when the plants are inoculated with *Pseudomonas chlororaphis* ST9.

3 Tolerance to abiotic stresses

Over the past years, several studies indicated that PGPB/PGPR inoculation can induce plant tolerance against different abiotic stresses (Alagna et al., 2020; Gamalero et al., 2020; Sangiorgio et al., 2020; Meena et al., 2017). Nevertheless, the level of tolerance depends on the microbial capability to induce the expression of stress-responsive transcription factors in plants as well as the production of enzymes involved in the detoxification of reactive oxygen species (ROS), synthesizing proline and EPS and biomass stabilization (Aeron et al., 2020).

Salt stress is one of the major threats to agriculture, negatively affecting crop yield and growth (Shrivastava and Kumar, 2015). It induces osmotic and ionic stress in plants, causing nutritional imbalance, morphological damages, less photosynthetic capacity and death (Ahmad et al., 2013). Unfortunately, high salinity areas are increasing every year, and agriculture has therefore to manage salt stress maintaining a sufficient crop production to satisfy food demand (Panwar et al., 2016). PGPR can alleviate the negative effects of salt by incrementing seed germination rate and leaf area, improving chlorophyll and protein content, increasing plant growth, productivity, and nutrient availability, delaying leaf senescence and enhancing tolerance to stresses

(Saghafi et al., 2019; Habib et al., 2016). PGPR ameliorate salt stress tolerance through several mechanisms, for example, accumulation of osmolytes operating in ion homeostasis, improvement of nutrient uptake (N, P, K, Zn and Si), production of ACC deaminase, indole acetic acid (IAA), sideropheres and EPS, and alteration of the antioxidant defence system (Saghafi et al., 2019 and reference therein). Upadhyay and Singh (2015) have demonstrated that salt-tolerant PGPR improved both growth and dry mass of wheat grown in pots, as well as root dry weight and shoot biomass in field conditions. Palaniyandi et al. (2014) demonstrated that inoculation of tomato plants with *Streptomyces* sp. strain PGPA39A under salt stress increased plant biomass and chlorophyll content, while leaf proline content decreased. In another study, it was shown that strains belonging to *Streptomyces* ameliorated salt stress tolerance in *Stevia* crops (Tolba et al., 2019). Panwar et al. (2016) suggested that using a combination of two PGPR (bacterial strains belonging to genus *Pantoea* and *Enterococcus*) on mung bean (*Vigna radiate*) plants resulted in enhanced growth and yield, a reduced Na+ concentration, less membrane damage and more antioxidants, such as ascorbic acid and glutathione, under salt stress. In the study of Khan et al. (2019), isolation and application of halotolerant PGPR on soybean plants grown under salt stress resulted in an increase in the antioxidant level, K+ uptake, plant growth attributes and chlorophyll content and a reduction of the Na+ ion concentration and the ABA level. Recently, Galicia-Campos et al. (2020) showed that the use of PGPR strains improved stress tolerance and water use efficiency in olive plants grown under saline stress.

Drought can also have a negative impact on crops causing significant yield reductions (Zhang et al., 2009; Breitkreuz et al., 2019). Many crops, including rice and winter wheat, need irrigation with big quantities of water in order to grow and produce acceptable yields. The research carried out by Zhang et al. (2020) showed that the association of rice roots with *Enterobacter aerogenes* is involved in rhizosheath (i.e. the layer of soil around the root containing a mixture of exudates, mucilage and exopolymers, which increases the wettability and water use efficiency of the root system) formation under moderate soil drying. It has been proposed that root-bacteria associations substantially contribute to this process by mechanisms that involve the ET response, considering that an ACC deaminase-deficient mutant of *E. aerogenes* failed to enhance rice rhizosheath formation. Breitkreuz et al. (2019) showed the positive role of *Phyllobacterium* in phosphate solubilization in rhizosphere under drought conditions. Brilli et al. (2019) demonstrated that tomato plants inoculated with *Pseudomonas chlororaphis* subsp. *aureofaciens* strain M71 have more proline and an improved antioxidant activity under mild water stress, thus reducing ROS presence and enhancing stress tolerance. The presence of the M71 strain also had an impact on stomatal closure, increasing ABA level in leaves and improving water use efficiency and biomass in water-stressed plants (Brilli et al., 2019). Rolli et al. (2015) demonstrated that PGPB have the ability to increase grapevine root biomass in field conditions under drought stress, while Saleem et al. (2018) showed that two PGPR strains improve velvet bean growth under drought conditions, by reducing ET production through ACC deaminase activity, which acts on the ET precursor ACC. Rubin et al. (2017) using a meta-analysis reported that PGPR can contribute to drought amelioration and water conservation, increasing shoot biomass and yield, especially under drought conditions.

Application of PGPR in combination with salicylic acid (SA) on maize plants (Khan et al., 2020) resulted in significant increases in nutrients content (Ca, K, Mg, Zn and Fe) in the shoots and the rhizosphere of plants and alleviated the adverse effects of low moisture stress of soil. Previously, Khan and Bano (2019) showed that the combination of PGPR and SA on wheat under drought stress led to a significant increase in leaf protein and sugar contents and higher chlorophyll content, chlorophyll fluorescence and performance index (Khan and Bano, 2019), suggesting the adoption of a mixed approach including both biological and chemical priming (Alagna et al., 2020).

In natural conditions, abiotic stresses can occur simultaneously, for example, salinity and phosphorous deficiency. It has been demonstrated that, under phosphate (Pi) limitation and salt stress, PGPR can support plant growth in plant genotype- and bacterial strain-dependent way (Tchuisseu Tchakounté et al., 2018). Osmotic stress and limitation of resources can also affect ornamental plants. It was shown that inoculation of petunia with *Pseudomonas* strains increased plant biomass and flowers number (Nordstedt et al., 2019). A study by Liu et al. (2019) focused on the physiological features and growth of North China red elder (*Sambucus williamsii*) under drought stress and in the presence of PGPR. *Acinetobacter calcoaceticus* X128 significantly increased stomatal conductance (Liu et al., 2019). The bacterium was able to increase cytokinins levels in the leaves that promote the stomatal opening, mitigating the inhibition of the photosynthetic rate in arid locations (Liu et al., 2019). In addition, the application of the PGPR strain might increase the permeability of roots to water or improve the transport of ions into the xylem, with an intensification of ABA transport, resulting in a decrease or complete closure in the stomatal opening (Liu et al., 2019). Generally, PGPR inoculation under drought conditions improved the adaptability of red elder plants to the arid environment by affecting phytohormones content in plants (Liu et al., 2019).

4 Beneficial effects against biotic stresses

The use of PGPB/PGPR is an eco-friendly tool that can be used for biocontrol of plant pathogens either by suppressing pathogenic microorganisms directly or by improving plant defence against pathogens (Lugtenberg and Kamilova, 2009). Controlling plant diseases by microorganisms is a complex process involving the biocontrol agent, the pathogen and the host, but also the indigenous microorganisms of the rhizosphere, other native macrobiota and the plant growth substrate. To act efficiently, the biocontrol microbe should remain active under varying conditions, such as temperature, moisture, pH and other soil properties.

Various mechanisms have been reported to be involved in biocontrol. The production of antibiotics and other antimicrobial metabolites is considered as a primary mechanism of biocontrol by PGPB and PGPR and is the most effective antagonistic activity to suppress phytopathogens. Diffusible antibiotics produced like phenazines, rhamnolipids, cyclic lipopeptides, zwittermycin A, kanosamine, oomycin A, ecomycins, butyrolactones and volatiles, such as hydrogen cyanide, ammonia, 2,3-butanediol and other blends of aldehydes, alcohols, ketones and sulphides, are known to possess antimicrobial and growth-promoting activities (Kai et al., 2009; Fernando et al, 2005). These compounds are toxic towards phytopathogens at concentrations depending on the compound and the target. Modes of action are not fully understood for many antimicrobial metabolites yet. In fungal pathogens, they may affect the cell membrane and zoospores (biosurfactants), inhibit the respiratory electron transport (phenazines, pyrrolnitrin) or cytochrome c oxidases and other metalloenzymes (hydrogen cyanide) (Raaijmakers, et al., 2006; Haas and Défago, 2005).

Another important mechanism in biocontrol is the production of hydrolytic enzymes by PGPB/PGPR directed against plant pathogens. Many biocontrol agents synthesize and secrete catabolic enzymes that can contribute to the suppression of phytopathogens through the hydrolysis of fungal cell wall components, such as cellulose, chitin, β-glucans and proteins (Abdullah et al., 2008; Dunne et al., 1997; Chernin et al., 1995). Production of β-1,3-glucanase by *Streptomyces* and *Paenibacillus* strains was shown to have an inhibitory effect on *F. oxysporum*, while *Bacillus cepacia* with glucanase activity showed inhibitory effect on many soilborne pathogens, including *Rhizoctonia solani*, *Sclerotium rolfsii* and *Pythium ultimum* (Compant et al., 2005). Several microorganisms were reported to show chitinolytic activity, including many *Bacillus*, *Streptomyces*, *Serratia* and *Pseudomonas* strains (Tsolakidou et al., 2019; Felse and Panda, 2000;). Co-cultivation of *Rhizoctonia solani* with the chitinolytic *Serratia marcescens* B2 strain led to several abnormalities of the mycelia (e.g. swelling, curling or bursting), suggesting degradation of the hyphal cell wall or hyphal cell death. Moreover, the application of *Serratia marcescens* B2 strain on cyclamen plants suppressed the diseases caused by *Rhizoctonia solani* and *F. oxysporum* f. sp. *cyclaminis* (Someya et al., 2000). Chitinases and cellulases are also involved in predation and parasitism, the major biocontrol mechanism used by *Trichoderma* and *Gliocladium* species (Harman et al., 2004a). This form of antagonism affects various fungal pathogens, such as *Sclerotinia*, *Rhizoctonia*, *Verticillium* and *Gaeumannomyces* (Harman et al., 2004b), and involves tropic growth of the BCA towards the target organism, coiling and dissolution of the pathogen's cell wall or membrane through enzymatic activity (Djonović et al., 2006; Woo et al., 2006; Zeilinger et al., 1999).

Apart from the mechanisms where a BCA produces substances with direct inhibitory effect for phytopathogens, it is possible for some PGPB/PGPR to outcompete the phytopathogens, either for space at the root surface or for nutrients, especially those secreted by the roots. This competition excludes pathogens by the physical occupation of binding sites on the root or by the depletion of food. Competition can take place for organic compounds necessary for pathogen proliferation and subsequent root colonization and for micronutrients that are essential for the growth and activity of the pathogen (Raaijmakers et al., 2009). Biocontrol based on competition for micronutrients has long been recognized, especially for nutrients that are not readily available for plants and microorganisms. Iron is a characteristic example of a micronutrient that is extremely limited in soils, and its availability depends on soil pH. In oxidized soils, iron is in the ferric form that is insoluble in water (Lindsay, 1979), and its concentration is too low to support the growth of microorganisms. To survive, microorganisms produce and secrete high-affinity chelators called siderophores (Neilands, 1995). Siderophore-producing PGPB/PGPR show increased efficiency in iron uptake making iron unavailable to pathogens and thus preventing their proliferation around the root, especially in soils with high pH (Kumar et al., 2015). Competition for iron as well as competition for carbon is an important mode of action of many biocontrol agents (Alabouvette et al., 2006; Lemanceau et al., 1992).

Besides functioning as BCAs, several PGPB and PGPR can induce a systemic response in the plant, leading to the activation of plant defence mechanisms against a wide range of phytopathogens (Pieterse et al., 2014). This form of resistance is referred to as induced systemic resistance (ISR) and is described as an enhanced defence capability of the plant against multiple pathogens (Conrath et al., 2015). ISR is induced by non-pathogenic PGPR, PGPB and fungi and can reduce the activity of pathogenic microorganisms via a complex system mediated by jasmonic acid (JA) and ET signalling (Pieterse et al., 2014; Van Loon, 1997). In contrast to classical biological control, in which the BCA is active against one or a few pathogens, ISR is effective against a broad spectrum of pathogens (Hariprasad et al., 2014). Several cell surface components and compounds produced by PGPR/PGPB can trigger ISR, including lipopolysaccharides and flagella, (Pieterse et al., 2003; Haas and Défago; 2005), siderophores (Meziane et al., 2005), volatiles (Ryu et al., 2004), hydrogen cyanide (HCN) (Defago et al., 1990), diacetylphloroglucinol (DAPG) (Weller et al., 2007) and cyclic lipopeptide surfactants (Ongena et al., 2007). The first reports of ISR were published back in 1991 and provided evidence that certain PGPR strains can stimulate the plant immune system and promote plant health (Alström, 1991; Van Peer et al., 1991; Wei et al., 1991). Since then, many studies have reported the ability of non-pathogenic microorganisms to trigger ISR including bacteria (e.g. *Pseudomonas, Serratia, Bacillus*), fungi (*F. oxysporum, Trichoderma, Piriformospora indica*) and symbiotic AMF (Pieterse et al., 2014). Enhancement in the plant's defence capability by ISR involves the activation of many biochemical pathways leading to fortification of structural barriers, such as thickened cell walls, suberization and deposition of lignin and callose (Raj et al., 2012; Benhamou et al., 1998). The phenomenon of ISR is also associated with increased expression of defence-related enzymes, such as phenylalanine ammonia lyases, peroxidases, lipoxygenases, polyphenol oxidases and synthesis of antimicrobial compounds, such as pathogenesis-related proteins, phytoalexins, phenolic compounds and cell wall peroxidases (Stringlis et al., 2018; van Loon et al., 1998; Zdor and Anderson, 1992; van Peer et al., 1991; Mauch et al., 1988).

5 Interaction between plant growth-promoting bacteria/rhizobacteria and arbuscular mycorrhizal fungi

The rhizosphere harbours a diverse community of microorganisms, such as bacteria and fungi that can interact with the plant, influencing plant growth, nutrition and health and protecting them from biotic and abiotic stresses in agro-ecosystems and in natural ecosystems (Philippot et al., 2013). AMF are one among the soilborne fungi that form symbiotic interactions with the majority of terrestrial plants. AMF are actively involved in the uptake of water and nutrients (such as phosphorus, nitrogen, zinc, copper, etc.) and increase resistance or tolerance of plants to biotic and abiotic stresses (Balestrini and Lumini, 2018).

AMF may interact synergistically with PGPR, leading to enhanced plant growth compared to single inoculation with either of them (Nanjundappa et al., 2019).

The review of Nanjundappa et al. (2019) focusing on the interaction between AMF and *Bacillus* concluded that combined inoculation leads to enhanced growth of plants, such as *Medicago sativa* (Medina et al., 2003), *Lactuca sativa* (Adriana et al., 2007), *Calendula officinalis* (Flores et al., 2007), *Artemisia annua* (Awasthi et al., 2011), *Pelargonium graveolens* (Alam et al., 2011), and *Cucumis sativus* (Rabab, 2014) as compared to single inoculation with either of them. Cely et al. (2016) also demonstrated that AMF and PGPR increased wood yield of *Schizolobium parahyba* var. *amazonicum* with respect to a fertilizer addition. Recently, Rocha et al. (2020) confirmed by field trials the positive role of co-presence of *Pseudomonas libanensis* and multiple AM fungal isolates of *Rhizophagus irregularis* in cowpea (*Vigna unguiculata*).

The plants showed increased shoot dry weight, pods and seeds per plant and grain yield (Rocha et al., 2020). In another study, a consortium of PGPR-rhizobia-AMF affected positively fava bean and wheat, improving shoot and root dry weight, leaf number, productivity and sugar, protein N, P, Ca, K and Na content (Raklami et al., 2019). Bona et al. (2015, 2017) demonstrated that the AMF-bacterium combined application can also affect fruit crop quality and nutritional value of strawberry and tomato (increased sugar content, fruit size, quantity and flowers) in conditions of reduced chemical inputs. The interaction between fungi and bacteria can also protect plants, by inducing systemic resistance to soilborne pathogens (Nanjundappa et al., 2019). For example, Jaizme-Vega et al. (2006) demonstrated a reduction of *Meloidogyne* infestation in AMF-PGPR-inoculated papaya plants, while Phirke et al. (2008) showed reduced *Fusarium* wilt in addition to improved yield in mycorrhized banana inoculated with rhizobacteria. The co-presence of AMF and PGPB/PGPR also improved tolerance to drought and salt stress in *Lactuca sativa* (Vivas et al., 2003), *Retama sphaerocarpa* (Marulanda et al., 46), *Z. mays* (Armada et al., 2015), *Trifolium repens* (Ortiz et al., 2015), *Lavandula dentate* (Armada et al., 2016) and *Acacia gerrardii* (Hashem et al., 2016). Recently, Inculet et al. (2019) demonstrated that inoculation of an irrigated tomato cultivar with AMF, PGPR and *Trichoderma*-based products increased plant length, fruit number, yield and quality traits based on lycopene and polyphenol content. Mannino et al. (2020), using different microbial inocula based on AM fungi or PGPR tolerant to salt, demonstrated that the tomato responses to water limitation depended on the inoculum composition. Balestrini et al. (2017) showed that the response of grapevine changed in the presence of a mixed inoculum composed by bacterial and fungal consortium compared to that with an inoculum based on *Funneliformis mosseae* only. Thus, the strategy of using a combination of AMF and PGPR in agricultural practice may improve soil health management, aiding nutrient solubilization and uptake and reduce the necessary fertilizer quantity. Nevertheless, more field studies are needed in order to verify the successful performance of the combined inoculations under real conditions (Nanjundappa et al., 2019).

The study of Todeschini et al. (2018) highlighted the importance of selecting the optimal combination of AMF and PGPR to positively influence physiological parameters, yield and quality in strawberry. The results of this study showed that application of the AMF affected the parameters associated with the plant vegetative portion, while application of the bacterium affected the fruit yield and quality. Interestingly, the volatile profile and elemental composition of the strawberry fruit were affected by the presence of a specific fungal–bacterial combination (Todeschini et al., 2018). This study showed for the first time that different soil microorganisms are able to influence the fruit concentration of some elements and/or volatiles (Todeschini et al., 2018).

Previous studies tested the ability of *Pseudomonas fluorescens* PGPR strains to form biofilm on mycorrhized and non- mycorrhized roots and on extraradical mycelium of an AM fungus (Bianciotto et al., 2001a, b). The nonmucoid wild-type strain *Pseudomonas fluorescens* CHA0 adhered very little on all surfaces, whereas two mucoid strains with increased production of acidic extracellular polysaccharides formed a dense and patchy bacterial layer on the roots and fungal structures (Bianciotto et al., 2001a). The results of this study suggest that increased adhesive properties of PGPR may lead to more stable interactions in mixed inocula and the rhizosphere. In another study, the bacterial components possibly involved in the attachment of two other PGPR (*Azospirillum* and *Rhizobium*) to AM roots and AM fungal structures were evaluated; mutants affected in EPS were tested in *in vitro* adhesion assays and shown to be strongly impaired in the attachment to both types of surfaces as well as to quartz fibres (Bianciotto et al., 2001b). Anchoring of PGPR to AMF seems to be a significant trait for a stable fungus–bacteria association that would improve the development of mixed inocula.

6 Conclusion and future trends in research

PGPB/PGPR can be promising economical and healthy alternatives to chemical fertilizers, antibiotics, herbicides, pesticides, with their abilities to improve agro-ecological sustainability. However, it is important to realize that PGPR showing a positive effect on a plant species may not have the same effect on others (Raj et al., 2020; Zeller et al., 2007). As explained by Timm et al. (2016), not all the microbes present in the soil have positive functions, so it is important to understand which microbial species should be employed to maximize plant growth, development and health (Xiao et al., 2020; Yuan et al., 2016; Mueller and Sachs, 2015). Recently, Finkel et al. (2020) demonstrated that a single bacterial genus in a complex microbiome modulates root growth. Interestingly, Guerrieri et al. (2020) suggested that using a consortium of native PGPR strains may represent a suitable solution in sustainable agriculture, to guarantee crop yield and quality, reducing the chemical input application. Apart from the studies on the efficacy of microbial inoculants on plants, their potential risks to other plants, animals, and humans must also be evaluated (Martínez-Hidalgo et al., 2019). Also, isolation, purification and characterization of microorganisms from saline habitats and inoculation of agricultural plants with them could be a successful strategy to increase tolerance and productivity of the plants grown under stress conditions (Saghafi et al., 2019). Escudero-Martinez and Bulgarelli (2019) highlighted that the genetic diversity of the crop microbiota is reduced compared to that of wild plants and that in combination with the application of human inputs, the agroecosystem resilience and sustainability to various stressors (e.g. climate change) is undermined. It is, therefore, desirable to carry out genetic mapping analyses, crossing interfertile wild and modern varieties, to discover host traits putatively influencing the recruitment and maintenance of the microbiota (Perez-Jaramillo et al., 2016; Schlaeppi and Bulgarelli, 2015). A concept named 'breeding for the plant microbiota' based on the development of plant varieties able to recruit specific microbial taxa may result in future crops that are less dependent on external inputs to produce acceptable yields (Escudero-Martinez and Bulgarelli, 2019; Bulgarelli et al., 2013; Wissuwa et al., 2009). Moreover, the prospect of using microbial mixtures as inoculants that can positively affect plant performance is gaining research interest. A substantial number of studies suggests that complex microbial consortia provide plants with increased growth and health as compared to single strains. However, our understanding of how members of microbial consortia interact with one another and with their hosts in nature is critical for the successful implementation of microbial synthetic communities (SynComs) with desired host outputs (Pascale et al., 2020; Tsolakidou et al., 2019). On the basis of these approaches, it will be possible to deal with challenges that agriculture shall meet in the coming years.

7 Acknowledgement

RB was supported by the CNR project FOE-2019 DBA.AD003.139.

8 Where to look for further information

8.1 Special issues on plant growth promoting rhizobacteria

- https://www.mdpi.com/journal/plants/special_issues/PGPB.
- https://www.sciencedirect.com/journal/microbiological-research/special-issue/10P22CLD85N.

8.2 Key research organizations

The Asian PGPR Society for Sustainable Agriculture (http://asianpgpr.com/index.php).

9 References

Abdullah, M. T., Ali, N. Y. and Suleman, P. (2008). Biological control of *Sclerotinia sclerotiorum* (Lib.) de Bary with *Trichoderma harzianum* and *Bacillus amyloliquefaciens*, *Crop Prot.* 27(10), 1354–1359.

Adriana, M. A., Rosario, A., Juan, R. L. and Ricardo, A. (2007). Differential effects of a *Bacillus megaterium* strain on *Lactuca sativa* plant growth depending on the origin of the arbuscular mycorrhizal fungus coinoculated: physiologic and biochemical traits, *J. Plant Growth Regul.* 27(10), 10–18.

Aeron, A., Khare, E., Jha, C. K., Meena, V. S., Aziz, S. M. A., Islam, M. T., Kim, K., Meena, S. K., Pattanayak, A., Rajashekara, H., Dubey, R. C., Maurya, B. R., Maheshwari, D. K., Saraf, M., Choudhary, M., Verma, R., Meena, H. N., Subbanna, A. R. N. S., Parihar, M., Shukla, S., Muthusamy, G., Bana, R. S., Bajpai, V. K., Han, Y. K., Rahman, M., Kumar, D., Singh, N. P. and Meena, R. K. (2020). Revisiting the plant growth-promoting rhizobacteria: lessons from the past and objectives for the future, *Arch. Microbiol.* 202(4), 665–676.

Ahmad, M., Zahir, Z. A., Khalid, M., Nazli, F. and Arshad, M. (2013). Efficacy of Rhizobium and *Pseudomonas* strains to improve physiology, ionic balance and quality of mung bean under salt-affected conditions on farmer's fields, *Plant Physiol. Biochem.* 63, 170–176.

Alabouvette, C., Olivain, C. and Steinberg, C. (2006). Biological control of plant diseases: the European situation, *Eur. J. Plant Pathol.* 114(3), 329–341.

Alagna, F., Balestrini, R., Chitarra, W., Marsico, A. D. and Nerva, L. (2020). Chapter 3– Getting ready with the priming: innovative weapons against biotic and abiotic crop enemies in a global changing scenario. In: Hossain, M. A., Liu, F., Burritt, D. J., Fujita, M. and Huang, B. (Eds), *Priming-Mediated Stress and Cross-Stress Tolerance in Crop Plants*, Academic Press, pp. 35–56. https://www.elsevier.com/books/priming-mediat ed-stress-and-cross-stress-tolerance-in-crop-plants/hossain/978-0-12-817892-8

Alam, M., Khaliq, A., Sattar, A., Shukla, R. S., Anwar, M. and Seema Dharni, S. (2011). Synergistic effect of arbuscular mycorrhizal fungi and *Bacillus subtilis* on the biomass and essential oil yield of rose-scented geranium (*Pelargonium graveolens*), *Arch. Agron. Soil Sci.* 57(8), 889–898.

Ali, S., Hameed, S., Shahid, M., Iqbal, M., Lazarovits, G. and Imran, A. (2020). Functional characterization of potential PGPR exhibiting broad-spectrum antifungal activity, *Microbiol. Res.* 232, 126389.

Alström, S. (1991). Induction of disease resistance in common bean susceptible to halo blight bacterial pathogen after seed bacterization with rhizosphere pseudomonads, *J. Gen. Appl. Microbiol.* 37(6), 495–501.

Ambrosini, A., Beneduzi, A., Stefanski, T., Pinheiro, F. G., Vargas, L. K. and Passaglia, L. M. P. (2012). Screening of plant growth promoting rhizobacteria isolated from sunflower (*Helianthus annuus* L.), *Plant Soil* 356(1–2), 245–264.

Andreozzi, A., Prieto, P., Mercado-Blanco, J., Monaco, S., Zampieri, E., Romano, S., Valè, G., Defez, R. and Bianco, C. (2019). Efficient colonization of the endophytes *Herbaspirillum huttiense* RCA24 and *Enterobacter cloacae* RCA25 influences the physiological parameters of *Oryza sativa* L. cv. Baldo rice, *Environ. Microbiol.* doi: 10.1111/1462-2920.14688.

Antoun, H. and Prévost, D. (2006). Ecology of plant growth promoting rhizobacteria. In: Siddiqui, Z. A. (Ed), *PGPR Biocontrol Biofertilization*, Springer, New York, pp. 1–38.

Armada, E., Azcón, R., López-Castillo, O. M., Calvo-Polanco, M. and Ruiz-Lozano, J. M. (2015). Autochthonous arbuscular mycorrhizal fungi and *Bacillus thuringiensis* from a degraded Mediterranean area can be used to improve physiological traits and performance of a plant of agronomic interest under drought conditions, *Plant Physiol. Biochem.* 90, 64–74.

Armada, E., Probanza, A., Roldán, A. and Azcón, R. (2016). Native plant growth promoting bacteria *Bacillus thuringiensis* and mixed or individual mycorrhizal species improved drought tolerance and oxidative metabolism in *Lavandula dentata* plants, *J. Plant Physiol.* 192, 1–12.

Arruda, L., Beneduzi, A., Martins, A., Lisboa, B., Lopes, C., Bertolo, F., Passaglia, L. M. P. and Vargas, L. K. (2013). Screening of rhizobacteria isolated from maize (*Zea mays* L.) in Rio Grande do Sul State (South Brazil) and analysis of their potential to improve plant growth, *Appl. Soil Ecol.* 63, 15-22.

Ashraf, M. A., Rasool, M. and Mirza, M. S. (2011). Nitrogen fixation and indole acetic acid production potential of bacteria isolated from rhizosphere of sugarcane (*Saccharum officinarum* L), *Adv. Biol. Res.* 5(6), 348-355.

Awan, S. A., Ilyas, N., Khan, I., Raza, M. A., Rehman, A. U., Rizwan, M., Rastogi, A., Tariq, R. and Brestic, M. (2020). *Bacillus siamensis* reduces cadmium accumulation and improves growth and antioxidant defense system in two wheat (*Triticum aestivum* L.) varieties, *Plants (Basel)* 9(7), 878.

Awasthi, A., Bharti, N., Nair, P., Singh, R., Shukla, A. K., Gupta, M. M., Darokar, M. P. and Kalra, A. (2011). Synergistic effect of *Glomus mosseae* and nitrogen fixing *Bacillus subtilis* strain Daz26 on artemisin content in *Artemisia annua* L., *Appl. Soil Ecol.* 49, 125-130.

Balestrini, R. and Lumini, E. (2018). Focus on mycorrhizal symbioses, *Appl. Soil Ecol.* 123, 299-304.

Balestrini, R., Salvioli, A., Dal Molin, A., Novero, M., Gabelli, G., Paparelli, E., Marroni, F. and Bonfante, P. (2017). Impact of an arbuscular mycorrhizal fungus versus a mixed microbial inoculum on the transcriptome reprogramming of grapevine roots, *Mycorrhiza* 27(5), 417-430.

Baritz, R., Wiese, L., Verbeke, I. and Vargas, R. (2018). Voluntary guidelines for sustainable soil management: global action for healthy soils. In: Ginzky, H., Dooley, E., Heuser, I., Kasimbazi, E., Markus, T. and Qin, T. (Eds), *International Yearbook of Soil Law and Policy* (vol. 2017), Springer, Cham, pp. 17-36.

Batista, B. D., Lacava, P. T., Ferrari, A., Teixeira-Silva, N. S., Bonatelli, M. L., Tsui, S., Mondin, M., Kitajima, E. W., Pereira, J. O., Azevedo, J. L. and Quecine, M. C. (2018). Screening of tropically derived, multi-trait plant growth- promoting rhizobacteria and evaluation of corn and soybean colonization ability, *Microbiol. Res.* 206, 33-42.

Bechtaoui, N., Raklami, A., Tahiri, A. I., Benidire, L., El Alaoui, A., Meddich, A., Göttfert, M. and Oufdou, K. (2019). Characterization of plant growth promoting rhizobacteria and their benefits on growth and phosphate nutrition of faba bean and wheat, *Biol. Open* 8(7), bio043968.

Benhamou, N., Kloepper, J. W. and Tuzun, S. (1998). Induction of resistance against *Fusarium* wilt of tomato by combination of chitosan with an endophytic bacterial strain: ultrastructure and cytochemistry of the host response, *Planta* 204(2), 153-168.

Bertani, I., Zampieri, E., Bez, C., Volante, A., Venturi, V. and Monaco, S. (2021). Isolation and Characterization of *Pseudomonas chlororaphis* Strain ST9; Rhizomicrobiota and in Planta Studies, *Plants* 10(7), 1466.

Bhattacharyya, D., Garladinne, M. and Lee, Y. H. (2015). Volatile indole produced by Rhizobacterium *Proteus vulgaris* JBLS202 stimulates growth of *Arabidopsis thaliana* through auxin, cytokinin, and brassinosteroid pathways, *J. Plant Growth Regul.* 34(1), 158-168.

Bhattacharyya, P. N. and Jha, D. K. (2012). Plant growth-promoting rhizobacteria (PGPR): emergence in agriculture, *World J. Microbiol. Biotechnol.* 28(4), 1327-1350.

Bianciotto, V., Andreotti, S., Balestrini, R., Bonfante, P. and Perotto, S. (2001a). Mucoid mutants of the biocontrol strain pseudomonas fluorescens CHA0 show increased ability in biofilm formation on mycorrhizal and nonmycorrhizal carrot roots, *Mol. Plant Microbe Interact.* 14(2), 255-260.

Bianciotto, V., Andreotti, S., Balestrini, R., Bonfante, P. and Perotto, S. (2001b). Extracellular polysaccharides are involved in the attachment of *Azospirillum brasilense* and *Rhizobium leguminosarum* to arbuscular mycorrhizal structures, *Eur. J. Histochem.* 45(1), 39–49.

Bisht, N., Tiwari, S., Singh, P. C., Niranjan, A. and Singh Chauhan, P. (2019). A multifaceted rhizobacterium *Paenibacillus lentimorbus* alleviates nutrient deficiency-induced stress in *Cicer arietinum* L., *Microbiol. Res.* 223–225, 110–119.

Bona, E., Cantamessa, S., Massa, N., Manassero, P., Marsano, F., Copetta, A., Lingua, G., D'Agostino, G., Gamalero, E. and Berta, G. (2017). Arbuscular mycorrhizal fungi and plant growth-promoting pseudomonads improve yield, quality and nutritional value of tomato: a field study, *Mycorrhiza* 27(1), 1–11.

Bona, E., Lingua, G., Manassero, P., Cantamessa, S., Marsano, F., Todeschini, V., Copetta, A., D'Agostino, G., Massa, N., Avidano, L., Gamalero, E. and Berta, G. (2015). AM fungi and PGP pseudomonads increase flowering, fruit production, and vitamin content in strawberry grown at low nitrogen and phosphorus levels, *Mycorrhiza* 25(3), 181–193.

Breitkreuz, C., Buscot, F., Tarkka, M. and Reitz, T. (2019). Shifts Between and Among populations of wheat rhizosphere *Pseudomonas*, *Streptomyces* and *Phyllobacterium* suggest consistent phosphate mobilization at different wheat growth stages Under abiotic stress, *Front. Microbiol.* 10, 3109.

Brilli, F., Pollastri, S., Raio, A., Baraldi, R., Neri, L., Bartolini, P., Podda, A., Loreto, F., Maserti, B. E. and Balestrini, R. (2019). Root colonization by *Pseudomonas chlororaphis* primes tomato (*Lycopersicum esculentum*) plants for enhanced tolerance to water stress, *J. Plant Physiol.* 232, 82–93.

Brodt, S., Six, J., Feenstra, G., Ingels, C. and Campbell, D. (2011). Sustainable agriculture, *Nat. Educ. Knowl.* 3(10), 1.

Bulgarelli, D., Schlaeppi, K., Spaepen, S., Ver Loren van Themaat, E. and Schulze-Lefert, P. (2013). Structure and functions of the bacterial microbiota of plants, *Annu. Rev. Plant Biol.* 64, 807–838.

Calvo, P., Watts, D. B., Kloepper, J. W. and Torbert, H. A. (2017). Effect of microbial-based inoculants on nutrient concentrations and early root morphology of corn (*Zea mays*), *J. Plant Nutr. Soil Sci.* 180(1), 56–70.

Cely, M. V. T., Siviero, M. A., Emiliano, J., Spago, F. R., Freitas, V. F., Barazetti, A. R., Goya, E. T., Lamberti, G. S., Dos Santos, I. M., De Oliveira, A. G. and Andrade, G. (2016). Inoculation of *Schizolobium parahyba* with mycorrhizal fungi and plant growth-promoting rhizobacteria increases wood yield under field conditions, *Front. Plant Sci.* 7, 1708.

Chernin, L., Ismailov, Z., Haran, S. and Chet, I. (1995). Chitinolytic *Enterobacter agglomerans* antagonistic to fungal plant pathogens, *Appl. Environ. Microbiol.* 61(5), 1720–1726.

Compant, S., Duffy, B., Nowak, J., Clement, C. and Barka, E. A. (2005). Use of plant growth-promoting bacteria for biocontrol of plant diseases: principles, mechanisms of action, and future prospects, *Appl. Environ. Microbiol.* 71(9), 4951–4959.

Compant, S., Samad, A., Faist, H. and Sessitsch, A. (2019). A review on the plant microbiome: ecology, functions, and emerging trends in microbial application, *J. Adv. Res.* 19, 29–37.

Conrath, U., Beckers, G. J., Langenbach, C. J. and Jaskiewicz, M. R. (2015). Priming for enhanced defense, *Annu. Rev. Phytopathol.* 53, 97–119.

Corwin, D. L. (2021). Climate change impacts on soil salinity in agricultural areas, *Eur. J. Soil Sci.* 72(2), 842–862. doi: 10.1111/ejss.13010.

Defago, G., Berling, C. H., Burger, U., Hass, D., Kahr, G., Keel, C., Voisard, C., Wirthner, P. and Wuthrich, B. (1990). Suppression of black root rot of tobacco and other root disease by strains of *Pseudomonas fluorescens*: potential applications and mechanisms. In: Schippers, D., Cook, R. J., Henis, Y., Ko, W. H., Rovira, A. D., Schippers, B. and Scott, P. R. (Eds), *Biological Control of Soil Borne Plant Pathogens*, CABI Publishing, Oxfordshire, UK, pp. 93–108.

Dimkpa, C., Weinand, T. and Asch, F. (2009). Plant–rhizobacteria interactions alleviate abiotic stress conditions, *Plant Cell Environ.* 32(12), 1682–1694.

Djonović, S., Pozo, M. J. and Kenerley, C. M. (2006). Tvbgn3 a b-16-glucanase from the biocontrol fungus *Trichoderma virens* is involved in mycoparasitism and control of *Pythium ultimum*, *Appl. Environ. Microbiol.* 72(12), 7661–7670.

dos Santos, R. M., Diaz, P. A. E., Lobo, L. L. B. and Rigobelo, E. C. (2020). Use of plant growth-promoting rhizobacteria in maize and sugarcane: characteristics and applications, *Front. Sustain. Food Syst.* 4, 136.

Dunne, C., Crowley, J. J., Moënne-Loccoz, Y., Dowling, D. N., Bruijn, S. and O'Gara, F. (1997). Biological control of *Pythium ultimum* by *Stenotrophomonas maltophilia* W81 is mediated by an extracellular proteolytic activity, *Microbiology (Reading)* 143(12), 3921–3931.

Escudero-Martinez, C. and Bulgarelli, D. (2019). Tracing the evolutionary routes of plant-microbiota interactions, *Curr. Opin. Microbiol.* 49, 34–40.

Etesami, H., Emami, S. and Alikhani, H. A. (2017). Potassium solubilizing bacteria (KSB): mechanisms, promotion of plant growth, and future prospects: a review, *J. Soil Sci. Plant Nutr.* 17(4), 897–911.

Farina, R., Beneduzi, A., Ambrosini, A., Campos, S. B., Lisboa, B. B., Wendisch, V., Vargas, L. K. and Passaglia, L. M. P. (2012). Diversity of plant growth-promoting rhizobacteria communities associated with the stages of canola growth, *Appl. Soil Ecol.* 55, 44–52.

Felse, P. A. and Panda, T. (2000). Production of microbial chitinases—a revisit, *Bioprocess Eng.* 23(2), 127–134.

Fernando, W. G. D., Nakkeeran, S. and Zhang, Y. (2005). Biosynthesis of antibiotics by PGPR and its relation in biocontrol of plant diseases. In: Siddiqui, Z. A. (Ed), *PGPR: Biocontrol and Biofertilization*, Springer, Dordrecht, pp. 67–109.

Finkel, O. M., Salas-González, I., Castrillo, G., Conway, J. M., Law, T. F., Teixeira, P. J. P. L., Wilson, E. D., Fitzpatrick, C. R., Jones, C. D. and Dangl, J. L. (2020). A single bacterial genus maintains root growth in a complex microbiome, *Nature* 587(7832), 103–108. doi: 10.1038/s41586-020-2778-7.

Flores, A. C., Luna, A. A. E. and Portugal, V. O. (2007). Yield and quality enhancement of marigold flowers by inoculation with *Bacillus subtilis* and *Glomus fasciculatum*, *J. Sustain. Agr.* 31(1), 21–31.

Galicia-Campos, E., Ramos-Solano, B., Montero-Palmero, M. B., Gutierrez-Mañero, F. J. and García-Villaraco, A. (2020). Management of plant physiology with beneficial bacteria to improve leaf bioactive profiles and plant adaptation under saline stress in *Olea europea* L, *Foods* 9(1), 57.

Gamalero, E., Bona, E., Todeschini, V. and Lingua, G. (2020). Saline and arid soils: impact on bacteria, plants, and their interaction, *Biology (Basel)* 9(6), 116.

Gamez, R. M., Rodríguez, F., Vidal, N. M., Ramirez, S., Vera Alvarez, R., Landsman, D. and Mariño-Ramírez, L. (2019). Banana (*Musa acuminata*) transcriptome profiling in response to rhizobacteria: *Bacillus amyloliquefaciens* Bs006 and *Pseudomonas fluorescens* Ps006, *BMC Genomics* 20(1), 378.

Gautam, S., Chauhan, A., Sharma, R., Sehgal, R. and Shirkot, C. K. (2019). Potential of *Bacillus amyloliquefaciens* for biocontrol of bacterial canker of tomato incited by *Clavibacter michiganensis* ssp. michiganensis, *Microb. Pathog.* 130, 196-203.

Glick, B. R. (2012). Plant growth-promoting bacteria: mechanisms and applications, *Scientifica (Cairo)* 2012, 963401.

Groover, W., Held, D., Lawrence, K. and Carson, K. (2020). Plant growth-promoting rhizobacteria: a novel management strategy for *Meloidogyne incognita* on turfgrass, *Pest Manag. Sci.* 76(9), 3127-3138. doi: 10.1002/ps.5867.

Guerrieri, M. C., Fanfoni, E., Fiorini, A., Trevisan, M. and Puglisi, E. (2020). Isolation and screening of extracellular PGPR from the rhizosphere of tomato plants after long-term reduced tillage and cover crops, *Plants (Basel)* 9(5), 668.

Haas, D. and Défago, G. (2005). Biological control of soil-borne pathogens by fluorescent pseudomonads, *Nat. Rev. Microbiol.* 3(4), 307-319.

Habib, S. H., Kausar, H. and Saud, H. M. (2016). Plant growth-promoting rhizobacteria enhance salinity stress tolerance in okra through ROS-scavenging enzymes, *BioMed Res. Int.* 2016, 6284547.

Hariprasad, P., Chandrashekar, S., Singh, S. B. and Niranjana, S. R. (2014). Mechanism of plant growth promotion and disease suppression by *Pseudomonas aeruginosa* strain 2apa, *J. Basic Microbiol.* 54(8), 792-801.

Harman, G. E., Howel, C. R., Viterbo, A., Chet, I. and Lorito, M. (2004a). *Trichoderma* species–opportunistic, avirulent plant symbionts, *Nat. Rev. Microbiol.* 2(1), 43-56.

Harman, G. E., Petzoldt, R., Comis, A. and Chen, J. (2004b). Interactions between *Trichoderma harzianum* strain T22 and maize inbred line Mo17 and effects of these interactions on diseases caused by *Pythium ultimum* and *Colletotrichum graminicola*, *Phytopathology* 94(2), 147-153.

Hashem, A., Abd_Allah, E. F., Alqarawi, A. A., Al-Huqail, A. A. and Shah, M. A. (2016). Induction of osmoregulation and modulation of salt stress in acacia gerrardii benth. by Arbuscular Mycorrhizal Fungi and Bacillus subtilis (BERA 71), *BioMed Res. Int.* 2016, 1-11, 6294098.

Heydari, M. M., Brook, R. M. and Jones, D. L. (2019). The role of phosphorus sources on root diameter, root length and root dry matter of barley (*Hordeum vulgare* L.), *J. Plant Nutr.* 42(1), 1-15.

Hyder, S., Gondal, A. S., Rizvi, Z. F., Ahmad, R., Alam, M. M., Hannan, A., Ahmed, W., Fatima, N. and Inam-Ul-Haq, M. (2020). Characterization of native plant growth promoting rhizobacteria and their anti-oomycete potential against *Phytophthora capsici* affecting chilli pepper (*Capsicum annum* L.), *Sci. Rep.* 10(1), 13859.

Inculet, C. S., Mihalache, G., Sellitto, V. M., Hlihor, R. M. and Stoleru, V. (2019). The effects of a microorganisms-based commercial product on the morphological, biochemical and yield of tomato plants under two different water regimes, *Microorganisms* 7(12), 706.

Jaizme-Vega, MdC., Rodriguez-Romero, A. S. and Barroso Nunez, L. A. (2006). Effect of the combined inoculation of arbuscular mycorrhizal fungi and plant growth promoting rhizobacteria on papaya (*Carica papaya* L.) infected with the root-knot nematode *Meloidogyne incognita*, *Fruits* 61(3), 151–162.

Jiang, C. H., Yao, X. F., Mi, D. D., Li, Z. J., Yang, B. Y., Zheng, Y., Qi, Y. J. and Guo, J. H. (2019). Comparative transcriptome analysis reveals the biocontrol mechanism of *Bacillus velezensis* F21 Against *Fusarium* Wilt on watermelon, *Front. Microbiol.* 10, 652.

Jiménez, J. A., Novinscak, A. and Filion, M. (2020). *Pseudomonas fluorescens* LBUM677 differentially increases plant biomass, total oil content and lipid composition in three oilseed crops, *J. Appl. Microbiol.* 128(4), 1119–1127.

Jiménez-Gómez, A., García-Estévez, I., García-Fraile, P., Escribano-Bailón, M. T. and Rivas, R. (2020). Increase in phenolic compounds of *Coriandrum sativum* L. after the application of a *Bacillus* halotolerans biofertilizer, *J. Sci. Food Agric.* 100(6), 2742–2749.

Kai, M., Haustein, M., Molina, F., Petri, A., Scholz, B. and Piechulla, B. (2009). Bacterial volatiles and their action potential, *Appl. Microbiol. Biotechnol.* 81(6), 1001–1012.

Kalam, S., Basu, A. and Podile, A. R. (2020). Functional and molecular characterization of plant growth promoting *Bacillus* isolates from tomato rhizosphere, *Heliyon* 6(8), e04734.

Khan, M. A., Asaf, S., Khan, A. L., Adhikari, A., Jan, R., Ali, S., Imran, M., Kim, K. M. and Lee, I. J. (2019). Halotolerant rhizobacterial strains mitigate the adverse effects of NaCl stress in soybean seedlings, *BioMed Res. Int.* 2019, 9530963.

Khan, N. and Bano, A. (2019). Exopolysaccharide producing rhizobacteria and their impact on growth and drought tolerance of wheat grown under rainfed conditions, *PLoS ONE* 14(9), e0222302.

Khan, N., Bano, A. and Curá, J. A. (2020). Role of beneficial microorganisms and salicylic acid in improving rainfed agriculture and future food safety, *Microorganisms* 8(7), 1018.

Khan, M. S., Zaidi, A., Ahemad, M., Oves, M. and Wani, P. A. (2010). Plant growth promotion by phosphate solubilizing fungi–current perspective, *Arch. Agron. Soil Sci.* 56(1), 73–98.

Khatoon, Z., Huang, S., Rafique, M., Fakhar, A., Kamran, M. A. and Santoyo, G. (2020). Unlocking the potential of plant growth-promoting rhizobacteria on soil health and the sustainability of agricultural systems, *J. Environ. Manage.* 273, 111118.

Kour, D., Rana, K. L., Yadav, N., Yadav, A. N., Kumar, A., Meena, V. S. and Saxena, A. K. (2019). Rhizospheric microbiomes: biodiversity, mechanisms of plant growth promotion, and biotechnological applications for sustainable agriculture. In: Kumar A., Meena V. (Eds), *Plant Growth Promoting Rhizobacteria for Agricultural Sustainability*, Springer, Singapore, pp. 19–65.

Kumar, A., Bahadur, I., Maurya, B. R., Raghuwanshi, R., Meena, V. S., Singh, D. K. and Dixit, J. (2015). Does a plant growth promoting rhizobacteria enhance agricultural sustainability?, *J. Pur. Appl. Microbiol.* 9, 715–724.

Lee, S., Trịnh, C. S., Lee, W. J., Jeong, C. Y., Truong, H. A., Chung, N., Kang, C. S. and Lee, H. (2020). *Bacillus subtilis* strain L1 promotes nitrate reductase activity in *Arabidopsis* and elicits enhanced growth performance in *Arabidopsis*, lettuce, and wheat, *J. Plant Res.* 133(2), 231–244.

Lemanceau, P., Bakker, P. A. H. M., de Kogel, W. J., Alabouvette, C. and Schippers, B. (1992). Effect of pseudobactin 358 production by *Pseudomonas putida* WCS358 on suppression of Fusarium wilt of carnation by nonpathogenic *Fusarium oxysporum* Fo47, *Appl. Environ. Microbiol.* 58(9), 2978-2982.

Lindsay, W. L. (1979). *Chemical Equilibria in Soils*, John Wiley & Sons, Inc., New York.

Liu, F., Ma, H., Du, Z., Ma, B., Liu, X. H., Peng, L. and Zhang, W. X. (2019). Physiological response of North China red elder container seedlings to inoculation with plant growth-promoting rhizobacteria under drought stress, *PLoS ONE* 14(12), e022 6624.

Lugtenberg, B. and Kamilova, F. (2009). Plant-growth-promoting rhizobacteria, *Annu. Rev. Microbiol.* 63, 541-556.

Ma, Y., Prasad, M. N., Rajkumar, M. and Freitas, H. (2011). Plant growth promoting rhizobacteria and endophytes accelerate phytoremediation of metalliferous soils, *Biotechnol. Adv.* 29(2), 248-258.

Mannino, G., Nerva, L., Gritli, T., Novero, M., Fiorilli, V., Bacem, M., Bertea, C. M., Lumini, E., Chitarra, W. and Balestrini, R. (2020). Effects of different microbial inocula on tomato tolerance to water deficit, *Agronomy* 10(2), 170.

Martínez-Hidalgo, P., Maymon, M., Pule-Meulenberg, F. and Hirsch, A. M. (2019). Engineering root microbiomes for healthier crops and soils using beneficial, environmentally safe bacteria, *Can. J. Microbiol.* 65(2), 91-104.

Mauch, F., Mauch-Mani, B. and Boller, T. (1988). Antifungal hydrolasses in pea tissue: II. Inhibition of fungal growth by combinations of chitinases and two β-1,3-glucanases, *Plant Physiol.* 87, 936-942.

Medina, A., Probanza, A., Gutierrez Mañero, F. J. and Azcón, R. (2003). Interactions of arbuscular-mycorrhizal fungi and *Bacillus* strains and their effects on plant growth, microbial rhizosphere activity (thymidine and leucine incorporation) and fungal biomass (ergosterol and chitin), *Appl. Soil Ecol.* 22(1), 15-28.

Meena, K. K., Sorty, A. M., Bitla, U. M., Choudhary, K., Gupta, P., Pareek, A., Singh, D. P., Prabha, R., Sahu, P. K., Gupta, V. K., Singh, H. B., Krishanani, K. K. and Minhas, P. S. (2017). Abiotic stress responses and microbe-mediated mitigation in plants: the omics strategies, *Front. Plant Sci.* 8, 172.

Meena, M., Swapnil, P., Divyanshu, K., Kumar, S., Harish, T., Tripathi, Y. N., Zehra, A., Marwal, A. and Upadhyay, R. S. (2020). PGPR-mediated induction of systemic resistance and physiochemical alterations in plants against the pathogens, current perspectives, *J. Basic Microbiol.* 60(10), 828-861.

Meziane, H., Van der Sluis, I., Van Loon, L. C., Hofte, M. and Bakker, P. A. (2005). Determinants of *Pseudomonas putida* WCS358 involved in inducing systemic resistance in plants, *Mol. Plant Pathol.* 6(2), 177-185.

Mueller, U. G. and Sachs, J. L. (2015). Engineering microbiomes to improve plant and animal health, *Trends Microbiol.* 23(10), 606-617.

Nanjundappa, A., Bagyaraj, D. J., Saxena, A. K., Kumar, M. and Chakdar, H. (2019). Interaction between arbuscular mycorrhizal fungi and *Bacillus* spp. in soil enhancing growth of crop plants, *Fungal Biol. Biotechnol.* 6, 23.

Neilands, J. B. (1995). Siderophores: structure and function of microbial iron transport compounds, *J. Biol. Chem.* 270(45), 26723-26726.

Niranjan Raj, S., Lavanya, S. N., Amruthesh, K. N., Niranjana, S. R., Reddy, M. S. and Shetty, H. S. (2012). Histo-chemical changes induced by PGPR during induction of resistance in pearl millet against downy mildew disease, *Biol. Control* 60(2), 90–102.

Nordstedt, N. P., Chapin, L. J., Taylor, C. G. and Jones, M. L. (2019). Identification of *Pseudomonas* spp. that increase ornamental crop quality during abiotic stress, *Front. Plant Sci.* 10, 1754.

Ongena, M., Jourdan, E., Adam, A., Paquot, M., Brans, A., Joris, B., Arpigny, J. L. and Thonart, P. (2007). Surfactin and fengycin lipopeptides of *Bacillus subtilis* as elicitors of induced systemic resistance in plants, *Environ. Microbiol.* 9(4), 1084–1090.

Orozco-Mosqueda, M. D. C., Rocha-Granados, M. D. C., Glick, B. R. and Santoyo, G. (2018). Microbiome engineering to improve biocontrol and plant growth-promoting mechanisms, *Microbiol. Res.* 208, 25–31.

Ortiz, N., Armada, E., Duque, E., Roldán, A. and Azcón, R. (2015). Contribution of arbuscular mycorrhizal fungi and/or bacteria to enhancing plant drought tolerance under natural soil conditions: effectiveness of autochthonous or allochthonous strains, *J. Plant Physiol.* 174, 87–96.

Palaniyandi, S. A., Damodharan, K., Yang, S. H. and Suh, J. W. (2014). *Streptomyces* sp. strain PGPA39 alleviates salt stress and promotes growth of "Micro Tom" tomato plants, *J. Appl. Microbiol.* 117(3), 766–773.

Panwar, M., Tewari, R. and Nayyar, H. (2016). Native halo-tolerant plant growth promoting rhizobacteria *Enterococcus* and *Pantoea* sp. improve seed yield of Mungbean (*Vigna radiata* L.) under soil salinity by reducing sodium uptake and stress injury, *Physiol. Mol. Biol. Plants* 22(4), 445–459.

Pascale, A., Proietti, S., Pantelides, I. S. and Stringlis, I. A. (2020). Modulation of the root microbiome by plant molecules: the basis for targeted disease suppression and plant growth promotion, *Front. Plant Sci.* 10, 1741.

Pellegrini, M., Ercole, C., Di Zio, C., Matteucci, F., Pace, L. and Del Gallo, M. (2020). *In vitro* and *in planta* antagonistic effects of plant growth-promoting rhizobacteria consortium against soilborne plant pathogens of *Solanum tuberosum* and *Solanum lycopersicum*, *FEMS Microbiol. Lett.* 367(13), fnaa099.

Pérez-Flores, P., Valencia-Cantero, E., Altamirano-Hernández, J., Pelagio-Flores, R., López-Bucio, J., García-Juárez, P. and Macías-Rodríguez, L. (2017). *Bacillus methylotrophicus* M4-96 isolated from maize (*Zea mays*) rhizoplane increases growth and auxin content in *Arabidopsis thaliana* via emission of volatiles, *Protoplasma* 254(6), 2201–2213.

Perez-Jaramillo, J. E., Mendes, R. and Raaijmakers, J. M. (2016). Impact of plant domestication on rhizosphere microbiome assembly and functions, *Plant Mol. Biol.* 90(6), 635–644.

Philippot, L., Raaijmakers, J. M., Lemanceau, P. and van der Putten, W. H. (2013). Going back to the roots: the microbial ecology of the rhizosphere, *Nat. Rev. Microbiol.* 11(11), 789–799.

Phirke, N. V., Kothari, R. M. and Chincholkar, S. B. (2008). Rhizobacteria in mycorrhizosphere improved plant health and yield of banana by offering proper nourishment and protection against diseases, *Appl. Biochem. Biotechnol.* 151(2–3), 441–451.

Pieterse, C. M. J., van Pelt, J. A., Verhagen, B. W. M., Ton, J., van Wees, S. C. M., Leon-Kloosterziel, K. M. and van Loon, L. C. (2003). Induced systemic resistance by plant growth-promoting rhizobacteria, *Symbiosis* 35, 39–54.

Pieterse, C. M. J., Zamioudis, C., Berendsen, R. L., Weller, D. M., Van Wees, S. C. M. and Bakker, P. A. H. M. (2014). Induced systemic resistance by beneficial microbes, *Annu. Rev. Phytopathol.* 52(1), 347–375.

Planchon, A., Durambur, G., Besnier, J. B., Plasson, C., Gügi, B., Bernard, S., Mérieau, A., Trouvé, J. P., Dubois, C., Laval, K., Driouich, A., Mollet, J. C. and Gattin, R. (2021). Effect of a *Bacillus subtilis* strain on flax protection against *Fusarium oxysporum* and its impact on the root and stem cell walls, *Plant Cell Environ.* 44(1), 304–322. doi: 10.1111/pce.13882.

Raaijmakers, J. M., de Bruijn, I. and de Kock, M. J. D. (2006). Cyclic lipopeptide production by plant-associated *Pseudomonas* species: diversity, activity, biosynthesis and regulation, *Mol. Plant-Microb. Interact.* 19, 699–710.

Raaijmakers, J. M., Paulitz, T. C., Steinberg, C., Alabouvette, C. and Moënne-Loccoz, Y. (2009). The rhizosphere: a playground and battlefield for soilborne pathogens and beneficial microorganisms, *Plant Soil* 321(1-2), 341–361.

Rabab, M. (2014). Interaction of *Bacillus subtilis* and *Trichoderma harzianum* with mycorrhiza on growth and yield of cucumber (*Cucumis sativus* L.), *Int. J. Curr. Res.* 6(8), 7754–7758.

Raj, M., Kumar, R., Lal, K., Sirisha, L., Chaudhary, R. and Kumar Patel, S. (2020). Dynamic role of plant growth promoting rhizobacteria (PGPR) in agriculture, *Int. J. Chemic, Stud* 8(5), 105–110.

Raklami, A., Bechtaoui, N., Tahiri, A. I., Anli, M., Meddich, A. and Oufdou, K. (2019). Use of rhizobacteria and mycorrhizae consortium in the open field as a strategy for improving crop nutrition, productivity and soil fertility, *Front. Microbiol.* 10, 1106.

Ravanbakhsh, M., Sasidharan, R., Voesenek, L. A. C. J., Kowalchuk, G. A. and Jousset, A. (2018). Microbial modulation of plant ethylene signaling: ecological and evolutionary consequences, *Microbiome* 6(1), 52.

Rocha, I., Souza-Alonso, P., Pereira, G., Ma, Y., Vosátka, M., Freitas, H.. and Oliveira, R. S. (2020). Using microbial seed coating for improving cowpea productivity under a low-input agricultural system, *J. Sci. Food Agric.* 100(3), 1092–1098.

Rolli, E., Marasco, R., Vigani, G., Ettoumi, B., Mapelli, F., Deangelis, M. L., Gandolfi, C., Casati, E., Previtali, F., Gerbino, R., Pierotti Cei, F., Borin, S., Sorlini, C., Zocchi, G. and Daffonchio, D. (2015). Improved plant resistance to drought is promoted by the root-associated microbiome as a water stress-dependent trait, *Environ. Microbiol.* 17(2), 316–331.

Rubin, R. L., van Groenigen, K. J. and Hungate, B. A. (2017). Plant growth promoting rhizobacteria are more effective under drought: a meta-analysis, *Plant Soil* 416(1-2), 309–323.

Ryu, C. M., Farag, M. A., Hu, C. H., Reddy, M. S., Kloepper, J. W. and Pare, P. W. (2004). Bacterial volatiles induce systemic resistance in *Arabidopsis*, *Plant Physiol.* 134(3), 1017–1026.

Saghafi, D., Delangiz, N., Lajayer, B. A. and Ghorbanpour, M. (2019). An overview on improvement of crop productivity in saline soils by halotolerant and halophilic PGPRs, *3 Biotech* 9(7), 261.

Saleem, A. R., Brunetti, C., Khalid, A., Della Rocca, G., Raio, A., Emiliani, G., De Carlo, A., Mahmood, T. and Centritto, M. (2018). Drought response of *Mucuna pruriens* (L.) DC. inoculated with ACC deaminase and IAA producing rhizobacteria, *PLoS ONE* 13(2), e0191218.

Sangiorgio, D., Cellini, A., Donati, I., Pastore, C., Onofrietti, C. and Spinelli, F. (2020). Facing climate change: application of microbial biostimulants to mitigate stress in horticultural crops, *Agronomy* 10(6), 794.

Santos, M. S., Nogueira, M. A. and Hungria, M. (2019). Microbial inoculants: reviewing the past, discussing the present and previewing an outstanding future for the use of beneficial bacteria in agriculture, *AMB Express* 9(1), 205.

Schlaeppi, K. and Bulgarelli, D. (2015). The plant microbiome at work, *Mol. Plant Microbe Interact.* 28(3), 212–217.

Sharma, R. K. and Archana, G. (2016). Cadmium minimization in food crops by cadmium resistant plant growth promoting rhizobacteria, *Appl. Soil Ecol.* 107, 66–78.

Shrivastava, P. and Kumar, R. (2015). Soil salinity: a serious environmental issue and plant growth promoting bacteria as one of the tools for its alleviation, *Saudi J. Biol. Sci.* 22(2), 123–131.

Sinha, A. K. and Parli, B. V. (2020). Siderophore production by bacteria isolated from mangrove sediments: a microcosm study, *J. Exp. Mar. Ecol.* 524, 151290.

Someya, N., Kataoka, N., Komagata, T., Hirayae, K., Hibi, T. and Akutsu, K. (2000). Biological control of cyclamen soilborne diseases by *Serratia marcescens* strain B2, *Plant Dis.* 84(3), 334–340.

Sonbarse, P. P., Kiran, K., Sharma, P. and Parvatam, G. (2020). Biochemical and molecular insights of PGPR application for the augmentation of carotenoids, tocopherols, and folate in the foliage of *Moringa oleifera*, *Phytochemistry* 179, 112506.

Stefan, M., Munteanu, N., Stoleru, V., Mihasan, M. and Hritcu, L. (2013). Seed inoculation with plant growth promoting rhizobacteria enhances photosynthesis and yield of runner bean (*Phaseolus coccineus* L.), *Sci. Hortic.* 151, 22–29.

Stringlis, I. A., Yu, K., Feussnerb, K., de Jonge, R., Van Bentum, S., Van Verk, M. C., Berendsen, R. L., Bakker, P. A. H. M., Feussner, I. and Pieterse, C. M. J. (2018). MYB72-dependent coumarin exudation shapes root microbiome assembly to promote plant health, *Proc. Natl. Acad. Sci. U. S. A.* 115(22), E5213–E5222.

Subbanna, A. R. N. S., Rajasekhara, H., Stanley, J., Mishra, K. K. and Pattanayak, A. (2018). Pesticidal prospectives of chitinolytic bacteria in agricultural pest management, *Soil Biol. Biochem.* 116, 52–66.

Tak, H. I., Ahmad, F. and Babalola, O. O. (2013). Advances in the application of plant growth-promoting rhizobacteria in phytoremediation of heavy metals, *Rev. Environ. Contam. Toxicol.* 223, 33–52.

Tchuisseu Tchakounté, G. V., Berger, B., Patz, S., Fankem, H. and Ruppel, S. (2018). Community structure and plant growth-promoting potential of cultivable bacteria isolated from Cameroon soil, *Microbiol. Res.* 214, 47–59.

Timm, C. M., Pelletier, D. A., Jawdy, S. S., Gunter, L. E., Henning, J. A., Engle, N., Aufrecht, J., Gee, E., Nookaew, I., Yang, Z., Lu, T. Y., Tschaplinski, T. J., Doktycz, M. J., Tuskan, G. A. and Weston, D. J. (2016). Two poplar-associated bacterial isolates induce additive favorable responses in a constructed plant-microbiome system, *Front. Plant Sci.* 7, 497.

Timmusk, S., Behers, L., Muthoni, J., Muraya, A. and Aronsson, A. C. (2017). Perspectives and challenges of microbial application for crop improvement, *Front. Plant Sci.* 8, 49.

Todeschini, V., AitLahmidi, N., Mazzucco, E., Marsano, F., Gosetti, F., Robotti, E., Bona, E., Massa, N., Bonneau, L., Marengo, E., Wipf, D., Berta, G. and Lingua, G. (2018). Impact of beneficial microorganisms on strawberry growth, fruit production, nutritional quality, and Volatilome, *Front. Plant Sci.* 9, 1611.

Tolba, S. T. M., Ibrahim, M., Amer, E. A. M. and Ahmed, D. A. M. (2019). First insights into salt tolerance improvement of *Stevia* by plant growth-promoting *Streptomyces* species, *Arch. Microbiol.* 201(9), 1295–1306.

Tsolakidou, M. D., Stringlis, I. A., Fanega-Sleziak, N., Papageorgiou, S., Tsalakou, A. and Pantelides, I. S. (2019). Rhizosphere-enriched microbes as a pool to design synthetic communities for reproducible beneficial outputs, *FEMS Microbiol. Ecol.* 95(10), fiz138.

Turan, M., Gulluce, M., von Wirén, N. and Sahin, F. (2012). Yield promotion and phosphorus solubilization by plant growth-promoting rhizobacteria in extensive wheat production in Turkey, *J. Plant Nutr. Soil Sci.* 175(6), 818–826.

Upadhyay, S. K. and Singh, D. P. (2015). Effect of salt-tolerant plant growth-promoting rhizobacteria on wheat plants and soil health in a saline environment, *Plant Biol. (Stuttg.)* 17(1), 288–293.

Van Loon, L. C. (1997). Induced resistance in plants and the role of pathogenesis-related proteins, *Eur. J. Plant Pathol.* 103(9), 753–765.

Van Loon, L. C., Bakker, P. A. H. M. and Pieterse, C. M. J. (1998). Systemic resistance induced by rhizosphere bacteria, *Annu. Rev. Phytopathol.* 36, 453–483.

Van Peer, R., Niemann, G. and Schippers, B. (1991). Induced resistance and phytoalexin accumulation in biological control of *Fusarium* wilt of carnation by *Pseudomonas* sp. strain WCS 417 r, *Phytopathology* 81(7), 728–734.

Vikram, A. and Hamzehzarghani, H. (2008). Effect of phosphate solubilizing bacteria on nodulation and growth parameters of greengram (*Vigna radiata* L. Wilczek), *Res. J. Microbiol.* 3(2), 62–72.

Vivas, A., Marulanda, A., Ruiz-Lozano, J. M., Barea, J. M. and Azcon, R. (2003). Influence of a *Bacillus* sp. on physiological activities of two arbuscular mycorrhizal fungi and on plant responses to PEG-induced drought stress, *Mycorrhiza* 13(5), 249–256.

Wei, G., Kloepper, J. W. and Tuzun, S. (1991). Induction of systemic resistance of cucumber to *Colletotrichum orbiculare* by select strains of plant growth-promoting rhizobacteria, *Phytopathology* 81(12), 1508–1512.

Weller, D. M., Landa, B. B., Mavrodi, O. V., Schroeder, K. L., De La Fuente, L., Bankhead, S. B., Molar, R. A., Bonsall, R. F., Mavrodi, D. M. and Thomashow, L. S. (2007). Role of 2,4-diacetylphloroglucinol producing fluorescent *Pseudomonas* spp. in plant defense, *Plant Biol.* 9, 4–20.

Wissuwa, M., Mazzola, M. and Picard, C. (2009). Novel approaches in plant breeding for rhizosphere-related traits, *Plant Soil* 321(1–2), 409–430.

Woo, S. L., Scala, F., Ruocco, M. and Lorito, M. (2006). The molecular biology of the interactions between *Trichoderma* spp. phytopathogenic fungi and plants, *Phytopathology* 96(2), 181–185.

Xiao, A. W., Li, Z., Li, W. C. and Ye, Z. H. (2020). The effect of plant growth-promoting rhizobacteria (PGPR) on arsenic accumulation and the growth of rice plants (*Oryza sativa* L.), *Chemosphere* 242, 125136.

Yuan, Z., Druzhinina, I. S., Labbé, J., Redman, R., Qin, Y., Rodriguez, R., Zhang, C., Tuskan, G. A. and Lin, F. (2016). Specialized microbiome of a halophyte and its role in helping non-host plants to withstand salinity, *Sci. Rep.* 6, 32467.

Zafar-Ul-Hye, M., Tahzeeb-Ul-Hassan, M., Abid, M., Fahad, S., Brtnicky, M., Dokulilova, T., Datta, R. and Danish, S. (2020). Potential role of compost mixed biochar with rhizobacteria in mitigating lead toxicity in spinach, *Sci. Rep.* 10(1), 12159.

Zaheer, A., Malik, A., Sher, A., Mansoor Qaisrani, M., Mehmood, A., Ullah Khan, S., Ashraf, M., Mirza, Z., Karim, S. and Rasool, M. (2019). Isolation, characterization, and effect of phosphate-zinc-solubilizing bacterial strains on chickpea (*Cicer arietinum* L.) growth, *Saudi J. Biol. Sci.* 26(5), 1061–1067.

Zdor, R. E. and Anderson, A. J. (1992). Influence of root colonizing bacteria on the defense responses of bean, *Plant Soil* 140(1), 99–107.

Zeilinger, S., Galhaup, C., Payer, K., Woo, S. L., Mach, R. L., Fekete, C., Lorito, M. and Kubicek, C. P. (1999). Chitinase gene expression during mycoparasitic interaction of *Trichoderma harzianum* with its host, *Fungal Genet. Biol.* 26(2), 131–140.

Zeller, S. L., Brandl, H. and Schmid, B. (2007). Host-plant selectivity of rhizobacteria in a crop/weed model system, *PLoS ONE* 2(9), e846.

Zhang, H., Xue, Y., Wang, Z., Yang, J. and Zhang, J. (2009). An alternate wetting and moderate soil drying regime improves root and shoot growth in rice, *Crop Sci.* 49(6), 2246–2260.

Zhang, Y., Du, H., Xu, F., Ding, Y., Gui, Y., Zhang, J. and Xu, W. (2020). Root-bacteria associations boost Rhizosheath Formation in moderately dry soil through ethylene responses, *Plant Physiol.* 183(2), 780–792.

www.ingramcontent.com/pod-product-compliance
Lightning Source LLC
Chambersburg PA
CBHW070720220326
41598CB00024BA/3240